Lecture Notes in Geoinformation and Cartography

Series Editors: William Cartwright, Georg Gartner, Liqiu Meng, Michael P. Peterson

For further volumes:
http://www.springer.com/series/7418

Sébastien Caquard • Laurene Vaughan
William Cartwright
Editors

Mapping Environmental Issues in the City

Arts and Cartography Cross Perspectives

Springer

Editors
Sébastien Caquard
Department of Geography,
Planning and Environment
Concordia University
Montréal Québec
Canada
scaquard@alcor.concordia.ca

William Cartwright
School of Mathematical and Geospatial Sciences
RMIT University
Melbourne Victoria
Australia
william.cartwright@rmit.edu.au

Laurene Vaughan
School of Media and Communication
RMIT University
Melbourne Victoria
Australia
laurene.vaughan@rmit.edu.au

ISSN 1863-2246 e-ISSN 1863-2351
ISBN 978-3-642-27114-4 ISBN 978-3-642-22441-6 (eBook)
DOI 10.1007/978-3-642-22441-6
Springer Heidelberg Dordrecht London New York

Softcover reprint of the hardcover 1st edition 2011

Printed on acid-free paper

Springer is part of Springer Science+Business Media (www.springer.com)

Preface

This book is the result of work from participants of the workshop entitled *'Mapping' Environmental Issues in the City: Arts and Cartography Cross-Perspective* that took place in Montréal in September 2010. During this workshop participants were invited to develop a mapping project inspired by a database compiled by the students of the Master of Environmental Impact Assessment (EIA) at Concordia University. We would like to thank all the students of the graduate course "GIS for Environmental Impact Assessment" of the winter term 2010, who contributed to this endeavor.

We also would like to thank the 18 reviewers who have contributed by their comments and suggestions to the refining of the chapters. Although most of the chapters went through a double blind peer review process, some of them did not - either because of timing issues or because their artistic approach did not really fit conventional peer review process. These chapters appear with the mention "Observations" in the book.

The workshop and the book project have received a financial support from Concordia University through the Aid to Research-Related Events program, as well as from the Department of Geography, Planning and Environment. Particularly, within this department we would like to thank Annie Pollock, department administrator, Aubyn O'Grady, undergraduate student, David Greene, chair of the department, and Undiné Thompson, graduate student, who undertook the task of the layout of the chapters. Finally, this project would not have been possible without the support of the International Cartographic Association (ICA), of the Royal Melbourne Institute of Technology (RMIT), and of the Mediatopias project at McGill University.

Sébastien Caquard, Montréal, Canada

Laurene Vaughan, Melbourne, Australia

William Cartwright Melbourne, Australia

Contents

Chapter 1

Mapping from Above/ Mapping from the Ground: Mapping Environmental Issues in the City

Sébastien Caquard[1]

Laurene Vaughan[2]

William Cartwright[3]

[1] Department of Geography, Planning and Environment, Concordia University, Montréal
[2] School of Media and Communication and Research Leader within the Design Research Institute at RMIT University, Melbourne
[3] School of Mathematical and Geospatial Sciences, RMIT University, Melbourne

Introduction

Expectations and conventions around "mapping" differ tremendously across different domains of knowledge and practice. While in cartography, mapping refers to specific techniques and concepts for representing information related to places, in the arts and humanities this terms is used more broadly and includes a diversity of practices and expectations. More than just data, artists and other creative practitioners express understandings, fears, hopes, emotions and perceptions about places and people through mapping, and these dimensions are essential for fully understanding our relation to places.

Contemporary artists have developed their own mapping practices as illustrated by recent publications such as *An Atlas of Radical Cartography* (Mogel & Bhagat 2008), *The Map as Art: Contemporary Artists Explore Cartography* (Harmon 2009), *Emotional Cartography Technologies of the Self* (Nold 2009), and *Journeys Beyond The Neatline – Expanding The Boundaries of Cartography* (Rangel et al. 2010). These artistic practices characterize a broad interest in the humanities for exploring the relationships between maps and artistic forms of expression such as painting (Orueta 2010), literature (Moretti 2007; Westphal 2007; Piatti et al. 2009) and cinema (Conley 2007; Castro 2009). The study of the relationships between arts and maps has also gained some interest within the discipline of cartography (Wood & Krygier 2006; Monmonier 2007; Cartwright 2010; Wood 2010).

The Art and Cartography working-group of the International Cartographic Association (ICA) was created in 2007 to stimulate the interaction between arts and cartography. Since then, this working group has organized its first symposium in 2008 in Vienna on "Cartography and Art - Art and Cartography", published an edited book (Cartwright et al. 2009), a special issue of *The Cartographic Journal* (Caquard et al. 2009) and organized a workshop entitled *"Mapping" Environmental Issues in the City: Arts and Cartographic Cross Perspectives,* in Montréal, Canada in September 2010. This workshop was designed to encourage and explore the interactions between cartographers, artists, designers and any other area of 'arts' who work in the various aspects of spatial representation, through the development of original mapping projects inspired by a common ground: a geographic database compiled to study environmental issues in the city of Montréal. This book formalizes the results of this workshop.

Contextualization

Database development is the first step of any map production (November et al. 2010). In the digital context, the role of databases in mapping processes is now paramount. According to new media theorist Lev Manovich (2000), the world, as an endless collection of images and files, needs to be approach as a database. This 'databasification' of the world reflects how computer structure defines the way the world must be categorized and analyzed (Leszczynski 2009). The production of the database creates a new world (Bowker 2000), and mapping the database becomes a way of making this world tangible. Although the map still serves to visualize "datascapes" as illustrated by Nadia Amoroso (2010), the relationship between maps and databases is changing dramatically.

As emphasized by Jeremy Crampton (2010a, p.4), in the discipline of cartography there is an "increasing emphasis on databases, rather than visual representation." The database is now what defines the world, while the map is more and more often envisioned as the interface to navigate through the growing wealth of data; it becomes what Valérie November and colleagues (2010, 583) call a "navigational platform." According to these authors, this "navigational platform" metaphor reflects the main function of the map in the growing volume of data. This idea is supported by the exponential use of maps by companies like Google to provide access to the wealth of data and information available online (Thielmann 2010). More and more often, Google returns maps as the result of searches done through its search engine.

Although there is an exponential development of user generated content in the context of Web 2.0 (Goodchild 2007; Crampton 2010b), compiling relevant and reliable databases remains a major challenge for environmental projects. As emphasized by Denis Wood (2010, p.164) in his critic of Participatory GIS projects, "the cost of collecting and organizing data about the environment is insane", which explains why institutions and governments remain the main sources of environmental data. Individuals, communities and organizations have often no other choice than to rely (at least partially) on these institutional data and on the way they frame environmental issues. What is changing is more the way maps are used to navigate through these data (as discussed previously), as well as to make environmental issues public and sometimes more tangible.

In a recent U.S. Senate Committee Hearing entitled, "Oversight Hearing on Disease Clusters and Environmental Health" (March 29, 2011), one of the testimonies was given by U.S. environmental and health advocate Erin Brockovich. She used a U.S. map locating all persons who contacted her to report health issues that might be related to pollution in their neighborhoods[1]. In her testimony the map becomes the tangible proof of the pervasive existence of health issues clusters related to the environment all over the U.S. This map serves both, as way to demonstrate the existence of clusters in terms of environmentally related health problems, and as a mean of communicating the existence of such issues. On this map, environmentally related health problems appear both clustered (locally) and spread (globally).

This map also provides a conventional view of phenomenon that are taking place on the ground. It synthesizes data while at the same time distancing itself from the individual dramas associated with each of these mapped cases. This is the way maps work according to cartographic principles. They provide an overview, by synthesizing information and erasing any associated emotions. They dissociate themselves from their object of study.

However, 'mapping' is quite different in the arts and humanities where it is often envisioned as a means of interacting with places in different ways. This dichotomy between the cartographic perspective on places from above, and the artistic perspective from the ground was apparent during the workshop. This dual perspective has provided the overall structure for this book. The first section presents the cartographic projects offering a perspective on environmental issues from above, while in the second section, the artistic projects map environmental issues from the ground.

[1] The full hearings and the map can be accessed on the U.S. senate web site: http://epw.senate.gov/public/index.cfm?FuseAction=Hearings.Hearing&Hearing_id=df802290-802a-23ad-480f-eba51b046c02

Mapping From Above: The Cartographic Perspectives

The goal of the workshop entitled *Mapping" Environmental Issues in the City: Arts and Cartographic Cross Perspectives* (Montréal, September 2010) was to provide the vehicle for bridging between internationally based artists, cartographers and individuals from the humanities interested in maps and mapping processes. About fifteen students, artists, cartographers, designers, and media practitioners (from Canada, USA, Europe, Australia, and New Zealand) were invited to attend the workshop, which was held over three days. The aim was to develop a project inspired by a given database. This database included data relative to health issues, socio-demographic profiles, air quality and industrial release in the city of Montréal. This database was compiled during the of winter 2010 by students undertaking the Master in Environmental Assessment (Department of Geography, Planning & Environment) at Concordia University (Montreal)[2]. The data collection and data analysis was part of a course entitled Geographic Information Systems for Environmental Impact Assessment (ENVS663). The goal of this course was to explore issues related to environmental injustices on the island of Montreal.

Undiné Thompson and Sébastien Caquard introduce and contextualize this database in chapter 2. This chapter emphasizes the pedagogical dimension of database production, as well as the limits inherent to the use of a geographic database (no matter how comprehensive) for addressing such a complex issue as environmental injustice. This contextualization chapter is followed by two chapters that explore the potential of some GIS tools and techniques for addressing specific environmental issues in Montréal. In chapter 3, Hiên Pham and colleagues use satellite imagery and statistical analysis to study the relationships between density of vegetation and sociodemographic profiles. Through this analysis they emphasize some relationships between low-income neighborhoods and limited access to vegetated areas, demonstrating the presence of few high inequity areas. In chapter 4, Tom Weatherburn and Daniel Naud analyze in depth the socio-demographic structure of Montréal through a geodemographic segmentation approach. This type of statistical analysis helps to refine the sociodemographic profiles of populations at risk of being subjects to environmental injustice. Throughout these three first chapters, it appears that the main sociodemographic divide in Montréal is more linguistic in nature than racial. The historical Francophone / Anglophone divide remains a major criteria for studying the geography of environmental issues in Montréal.

Chapter 5, 6 and 7 are developed by cartographers and illustrate the diversity of contemporary cartographic practices to represent environmental data. In chapter 5, Sven Fuhrmann proposes to rematerialize the health map by replacing conventional

[2] The database is available online (http://mappingworkshop.wordpress.com/database/) in different formats (e.g. excel spreadsheets, ArcGIS shapefiles, KMZ files and JPEG images).

proportional symbols with real prescription bottles of different sizes and colors. Fuhrmann argues that the use of tangible objects, such as prescription bottles, to map health issues could help improving public health communication and awarness. In chapter 6, Glenn Brauen focuses on the use of audio-visual mapping to represent air quality and, more specifically, BTEX concentration. The online interactive audiovisual map developed by Brauen is one of the very few cartographic examples using sound not as a redundant or mimetic variable but as a way to make the otherwise invisible 'visible' as audible outputs. Air pollution is invisible, whilst being pervasive and fluctuating. This map conveys in a very subtle way these different dimensions and stimuates the development of audiovisual mapping into new directions. In the last chapter of this first section, William Cartwright proposes to repurpose the MetroMap concept in the context of Web 2.0. In this chapter, Cartwright presents an online MetroMap prototype that uses real-time geo-located feeds to annotate a map of Montréal. Information mapped concerns four domains: environment, social & cultural, economic, and governance. This application cut the city in different slices: thematic slices, spatial slices and temporal slices. The idea of the city being represented as an assemblage of slices resonates with the concept of *assemblage* discussed by Kathy Waghorn in the second section of the book.

Mapping From The Ground: The Artistic/Humanistic Perspectives

In the second section of the book, artists and individuals from the humanities propose different ways of mapping environmental issues in the city from the ground. The database provided served mainly as a guide to orient the exploration of environmental issues in the city. It provides a point of departure for various modes of drifting and interacting with the city.

In the first chapter of this second section (Chapter 8), Laurene Vaughan provides a personal perspective on what a database might be. Through her *Roaming Montréal* project, Vaughan challenges the notion of abstract/objective databases, as well as the cartographic saying that maps are the best way to represent complex data and to communicate information associated to places. She then provides a more fluid personal, emotional and experiential sense to what a database could be through a walking activity. What a database actually is is definitely not something about which we all agree.

In chapter 9, Sarah Kanouse revisits the figure of the naturalist through the presentation of a "post-naturalist field-kit" developed to study "post-natural urban landscapes." This beautifully designed kit contains different tools and items to assist field studies. This project is deeply rooted in historical naturalistic approaches and design practices, as well as in social ecology and community-based activity. By making simple tools available to community members in order for them to collect and produce their own pieces of evidence of the environmental situation in their neighborhoods, Kanouse's approach definitely resonates with the geographical expeditions led by Bill Bunge in Detroit in the 1970s[3], and supports the idea and the importance of user-generated environmental data.

The importance of combining both the objective cartographic approach, with the more personal artistic perspective to address complex spatial phenomena is developed by Annalise Rees in chapter 10. Through the description of her drawing practices, Rees reflects on the importance of presence and perception in our relationships with places, and in mapping activities. In chapter 11, Kathy Waghorn envisions the city through Manuel De Landa's assemblage theory, which allows one to avoid "the reductive 'master narrative', so often activated as the 'master plan' in urban design terms." Building on the work of architect James Corner she envisions mapping as a "creative activity" determined more by its process than by its outcome. This perspective is materialized throughout her Montreal Garden Mapping project in which she proposes an assemblage illustrating the complexity and the diversities at stake in urban gardening. From this perspective, mapping provides the means to reveal complexity, instead of a way of simplifying it.

In chapter 12, John Calvelli proposes the application of his new photographic practice called "correlative ontography' – for mapping selected neighborhoods, based on environmental criteria provided via the database. The pictures taken during this process are then associated with different data characterizing the areas photographed. Through this process, Calvelli emphasizes the opposition between the pictures and the data, the personal and the measured, the visual discourse and the scientific.

Chapter 13 and 14 provide two different perspectives on sounds in the city. In chapter 13, Maryclare Foa proposes to improve our understanding of place through raising our awareness of ambient sound in the city. This is done through collaborative Driftsong performances, leading participants to react to ambient sounds, either through displacement or through the production of 'reacting' sounds. Foa considers this method - inspired from other drifting practices (such as the situationnist's *dérive*) – as being "the most democratic method of mapping place" because of its inherent lack of precision. Here again, artistic mapping practices serve to challenge conventional cartographic perspectives. In chapter 14, Jordan Lacey and Lawrence Harvey build on previous 'soundscape' work, proposing relevant locations for sites-of-respite in downtown Melbourne, Australia. These sites are envisioned both as places where one might take sound breaks from the noisy city, as well as indicating potential "performance locations for 'soundscape composition'.

3 See Wood (2010) for a recent review of Bunge's work.

Conclusion

Although the main goal of the workshop was to support interaction between cartographers and artists, a major dichotomy became quickly apparent in the way cartographers and artists approached the idea of mapping a common database as discussed in the conclusive chapter of this volume. While the former see the database as a source of *information* that can be mapped from above, the latter envisioned it as a source of *inspiration* to orient their research efforts from the ground. The map can serve as a way to navigate through the database, and to visualize its structure, just as the database can serve as a way to navigate through the world. This dichotomy emphasizes the complementary elements that exist between cartographic and artistic practices in their effort of portraying the world through mapping (in its widest context). From an artistic perspective mapping is personal, subjective, associated to ground exploration and dedicated to conveying the complexity of the phenomena being studied. From a cartographic perspective, mapping is data dependant, as objective as possible, provides a view from above distanced form the phenomena mapped, and largely dedicated to simplify the complexity of the phenomena represented as much as possible. Both of these approaches, and many others, are required for tracing environmental issues; as a way of leaving traces of their existence on maps, as well as tracing back their underlying causes and eventual consequences.

References

Amoroso, N., 2010. *The exposed city: mapping the urban invisibles*, Taylor & Francis.

Bowker, G.C., 2000. Biodiversity datadiversity. *Social Studies of Science*, 30, 643-683.

Caquard, S., Piatti, B. & Cartwright, W., 2009. Special Issue on Art & Cartography. *The Cartographic Journal*, 46(4), 289-291.

Cartwright, W., 2010. Mapping Other (Geographical) Realities. In *Mapping Different Geographies*. K. Kriz, W. Cartwright and L. Hurni (Eds.), Springer, Berlin Heidelberg, 11-35.

Cartwright, W., Gartner, G. & Lehn, A., 2009. *Cartography and art*, Springer, Berlin Heidelberg.

Castro, T., 2009. Cinema's Mapping Impulse: Questioning Visual Culture. *The Cartographic Journal*, 46(1), 9-15.

Conley, T., 2007. *Cartographic Cinema*, U of Minnesota Press, Minneapolis.

Crampton, J.W., 2010a. Cartographic calculations of territory. *Progress in Human Geography*. 35(1), 92-103.

Crampton, J.W., 2010b. *Mapping: A Critical Introduction to Cartography and GIS*, Oxford, UK: Wiley-Blackwell.

Goodchild, M.F., 2007. Citizens as sensors: the world of volunteered geography. *GeoJournal*, 69(4), 211-221.

Harmon, K., 2009. *The Map as Art: Contemporary Artists Explore Cartography*, Architectural

Press, Princeton.

Leszczynski, A., 2009. Rematerializing GIScience. *Environment and Planning D: Society and Space*, 27(4), 609-615.

Manovich, L., 2000. Database as a Genre of New Media. *AI & Society*, 14, 176-183.

Mogel, L. & Bhagat, A., 2008. *An Atlas of Radical cartography*, Journal of Aesthetics and Protest Press., Los Angeles.

Monmonier, M., 2007. Cartography: the multidisciplinary pluralism of cartographic art, geospatial technology, and empirical scholarship. *Progress in Human Geography*, 31(3), 371-379.

Moretti, F., 2007. *Graphs, Maps, Trees: Abstract Models for Literary History*, Verso.

Nold, C., 2009. *Emotional cartography. Technologies of the Self*, Retrieved May 2011, from www.emotionalcartography.net.

November, V., Camacho-Hübner, E. & Latour, B., 2010. Entering a risky territory: space in the age of digital navigation. *Environment and Planning D: Society and Space*, 28(4), 581-599.

Orueta, A.G., 2010. La cartografÍa en el cine: mapas y planos en las producciones cinemato-grÁficas occidentales. *Scripta Nova*, XIV(334). Retrieved March 2011, from http://www.ub.edu/geocrit/sn/sn-334.htm.

Piatti, B., Bär, H., Reuschel, A., Hurni, L., & Cartwright, W., 2009. Mapping Literature: Towards a Geography of Fiction. In *Cartography and Art*. W. Cartwright, G. Gartner, A. Lehn (Eds.), Springer, Berlin Heidelberg, 177-192.

Rangel, M.J., Coulis, M.J. & Jones, D.L., 2010. *Expanding the Boundaries of Cartography. Journeys Beyond the Neatline.* The University of Alberta Press., Edmonton.

Thielmann, T., 2010. Locative Media and Mediated Localities. *Aether: The Journal of Media Geography*, 5a, 1-17.

Westphal, B., 2007. *La Géocritique: Réel, Fiction, Espace*. Éditions de Minuit, Paris.

Wood, D., 2010. *Rethinking the Power of Maps*, The Guilford Press.

Wood, D. & Krygier, J., 2006. Art and Mapping: Special Issue. *Cartographic Perspectives*, 53.

Section I

Mapping From Above: The Cartographic Perspective

Chapter 2

Compiling a Geographic Database to Study Environmental Injustice in Montréal: Process, Results, and Lessons

Undiné Thompson and Sébastien Caquard

Department of Geography, Planning & Environment, Concordia University, Montréal

ABSTRACT

Environmental injustice is a concept that emerged out of the social impacts of environmental degradation within the United States during the early 1980s. In the U.S., environmental injustice is frequently tied to race while in Canada and internationally, more attention has been paid to the correlations between income and environmental hazards. This chapter presents a pedagogical project that aimed to assess the degree and structure of environmental injustice in the city of Montréal, Canada. To reach this goal, a group of graduate students in a course entitled "GIS for Environmental Impact Assessment" (ENVS 663) at Concordia University, Montréal, gathered data on a large variety of socio-economic, health and environmental factors on the Island of Montréal and then attempted to spatially analyze whether there are local areas that suffer heightened risk for environmental injustices. Several neighbourhoods were noted as at risk in these studies, which points to a need for greater research and investigation into the levels of disparity in Montréal and the impacts that pollutants may have on the Island's most at-risk residents. Through this process students were also exposed to the multiples problems associated with the development of a comprehensive and relevant database in order to study complex environmental issues. Students were able to overcome some of these problems and to collectively compile a database that served as a point of departure for organizing an international workshop entitled *'Mapping' Environmental Issues in the City: Arts and Cartography Cross-Perspectives*.

2.1 Introduction

Economic progress of society has long been associated with the creation of environmental "bads", or, as Aldo Leopold coined them, "regrettable necessities", that have the capacity to cause environmental problems, health issues and even human death (as cited by Buzzelli, 2008). Normally, pollutants are not evenly distributed in the air, soil, and water, and this inequality of distribution has frequently been found to place underprivileged social groups at heightened risk of exposure. This, in turn, has given rise to environmental injustices, which are defined as "the disproportionate exposure of communities of colour and the poor to pollution, and its concomitant effects on health and environment" (Maantay 2002, p. 161). This concept of environmental injustice also includes the more specific idea of environmental racism that argues that minority groups and aboriginal people tend to be at a greater risk of being affected by environmental hazards (Bullard 1990; Westra 2008). This occurs when specific segments of the population with heightened levels of deprivation suffer higher than normal morbidity and mortality statistics due to environmental effects[1]. These heightened levels of disease and/or death are often connected to the inequitable distribution of environmental health burdens associated with societal and economic development, and can include issues such as: pesticide use, air pollution, children's blood lead levels, toxins in food, hazardous waste facilities, hazardous releases (air and groundwater) and facility siting (White 1998).

Linked with this concept of environmental injustice is its more idealistic counterpart of environmental justice. Environmental justice is defined as the fair treatment and meaningful involvement of all people "regardless of race, colour, national origin, or income with respect to the development, implementation, and enforcement of environmental laws, regulations, and policies" (U.S. E.P.A., 2010).This chapter aims to enhance the greater understanding of the topic of environmental injustice through the development of a geospatial database. It will begin with a brief outline of the roots and evolution of environmental injustice as a concept, as well as its importance in Canada, and most significantly, Montréal. This chapter will then briefly present the history and geography of Montréal and explore previous local studies of environmental injustice. Subsequently the project of developing a geographic database on environmental injustice in Montréal that was undertaken by the ENVS 663 geographic information sciences class at Concordia University will be described. Following, this chapter will move to a discussion on the findings of this research and the challenges that existed. Overall, this chapter presents the development of the database that served as the reference for the participants of the workshop that was organized in Montréal in September 2011.

[1] In this context deprivation is defined as a set of circumstances that cause a person, family or group to be at an "observable and demonstrable disadvantage" in comparison to the local community, the larger society or even the national or global community (Townsend 1987, p. 125).

2.2 Environmental Injustice

2.2.1 Background

The concept of environmental injustice first emerged in the United States in the early 1980s when researchers such as Robert Bullard and Harvey White began investigating environmental decision making and policies that reflect the dominant power regimes in society and its institutions[2]. Researchers noted a social gradient in health levels, and, more specifically, that people that belonging to racial minorities and/or socioeconomically disadvantaged groups tend towards having inferior health when compared to their more affluent and advantaged counterparts (Pampalon and Raymond, 2002). Indeed, by the early 1990s an abundance of research had emerged that showed that Blacks, socio-economically disadvantaged, and working-class individuals were being subjected to disproportionately large amounts of pollution and other environmental stressors (externalities) in their neighbourhoods and their workplaces (Bullard, 1990).

The idea of environmental injustice continued to spread, and more research began to emerge through to the late 1990s and into the new century, moving beyond the American context and to a broader scope with attempts to understand the why "natural" events such as disease, are in reality, not natural at all but social (Robbins, 2005). Social scientists describe a "multidimensional web" of contextual forces and factors that create disproportionate vulnerability between and within neighbourhoods, cities, and countries (Bullard, 1990, p. 5; Robbins, 2005). Commonly cited risk-factors include occupation, education, value and type of land and dwelling, source and amount of income, government and private industry policies, and the racial and ethnic makeup of residents (Bullard, 1990; White, 1998).

The primary causes of environmental injustices include institutionalized racism; the co-modification of land, water, energy and air; unresponsive, unaccountable government policies and regulation; and lack of resources and power in affected communities (Ross, 2004). Proponents of environmentally degrading projects are rarely exposed to the externalities produced, nor are they usually required to take responsibility for their creation, while, in contrast, the less- wealthy, empowered, or dominant groups are frequently required to bear the burden of environmental degradation and associated negative health implications. Addressing the question of who profits from and who pays for "current environmental and industrial policies" is central within the analysis of environmental *in*/justice as well as the concomitant relationships between the dominant/affluent and the exploited segments of the population (Bullard, 1990, p. 98).

2.2.2 Environmental Injustice in Canada

Environmental injustice studies in Canada do not have as long a history as those in the United States, nevertheless, the "track record is small but growing" (Buzzelli *et al.*, 2003, p. 7). To date, the majority of the research in Canada has been concerned with environmental injustices related to air quality and attempting to understand the relationship between life expectancy and income (see Buzzelli *et al.*, 2003; Gower *et al.*, 2008; Crouse *et al.*, 2009).

Separate studies in cities such as Hamilton, Toronto and Montréal have found that there was air quality environmental injustice in each of these locations, but did not concur on the demographic groups that were at risk in each situation. In studies that were concurrently published in 2002, Ross and colleagues as well as Lobmayer and Wilkinson reported that there was a correlation between income levels and segregation in the United States, but that this was not the case in Canada. Since then the connection between income and life-expectancy and mortality in Canada has been further explored and it has been found that this relationship is not a simply linear, but rather that increased life expectancies were not related to higher income levels (Ross, 2008).

Despite this lack of a strong correlation between income and health levels in Canada, studies by Buzelli et al. (2003), Ross et al. (2003/2004), and Buzzelli and Jerret (2007) have shown that heightened levels of air pollutants exist in regions of lower socio-economic standing which tend to be in close proximity to major transportation corridors. These include areas where there are increased levels of single parent households and low education levels (Hamilton); where there are specific groups of visible minorities including Latinos, though the opposite observations were found for Black and Korean communities (Toronto); and where there are increased amounts of single-parent families with lower education levels and *contrary to previous research*, low incomes (Toronto). Canadian Aboriginal populations are commonly exposed to greater levels of environmental hazards than proximal non-aboriginal communities (Brody, 1997; Westra, 2008). There has also been research into the levels of environmental injustice in Montréal, but in preparation for exploring its particular Canadian nuances at the local level, this chapter will first briefly revisit Montréal's industrial history, political boundaries and social geography.

2.2.3 Montréal's Geography and Environmental Injustice

Montréal, Québec, is the second largest city in Canada, with 3.6 million people (Statistics Canada, 2006). The core of the city is located on a large island on the Saint-Lawrence River[2]. The history of the city is based upon its location as a trade

[2] The Island of Montréal has a population of just below two million inhabitants.

hub at the convergence of the Atlantic Ocean and the Great Lakes. Industrialization of Montréal began in earnest as it emerged as a key transportation link for North American trade over two hundred years ago upon the development of train and rail infrastructure and the Lachine Canal (1821), the forerunner of the St. Lawrence Seaway. This emphasis on Montréal as a key junction for local, national and international trade and shipping (Statistics Canada, 2006) has impelled the city's transition to a modern metropolis. Although this development has been a boon for the economy of Montréal, Québec and Canada, it has also been accompanied by increasing levels of pollution from the industries that support vehicles[3], these vehicles themselves[4], and from the industry that they support. Specifically, two major sources of pollutants in Montréal are the two oil refineries that exist on the east side of the island (Note: one was decommissioned in late 2010). There are numerous other point sources for pollutants in Montréal, and they are also primarily located in the eastern region of the island, as well as many linear road sources of air pollutants, as multiple highways criss-cross the island.

The social geography of the island has seen the proliferation of socially and materially advantaged inner cities on the slopes of Mount Royal since the 1800s with the language spoken in the home being the key point of differentiation within Montréal's neighbourhoods and boroughs. Historically the economic upper classes have been dominated by old Montréal-bred English-speaking families, with a modicum of French and others in the old-inner suburbs of Mont Royal, Outremont and Westmount (Ley, 1993). The middle classes are more mixed, whereas the lower economic stratum continues to be made up mainly of French and Irish Canadians and new immigrants (Buzzelli, 2008). The majority of the Francophone districts are located in the east of the island and in the southwest of the downtown core in the former industrial area near Lachine Canal. This includes regions like Pointe-Saint-Charles, Saint-Henri, Lachine, as well as other neighbourhoods such as Mercier, Hochelaga-Maisonneuve, Parc Extension and Montréal North which are the most economically deprived on the Island of Montréal (Langlois and Kitchen, 2001) (see Figure 2.1).

It has long been the case that the far west of the island has been primarily composed of wealthier Anglophone communities. Furthermore, there has been a recent turn towards the gentrification of lower class neighbourhoods in the south and east, like Saint-Henri, Pointe-Saint-Charles and Le Plateau (Ley, 1993). Although there are many discrete cultural neighbourhoods on the Island of Montréal as well as several regions of concentrated poverty, there are no true ethnic ghettos, nor ghettos of socially deprived groups as can be observed in large cities within the United States (Hatfield, 1997; Walks and Bourne, 2006). Additionally, there is no

[3] Including the petroleum industry and the creation and maintenance of the infrastructure.
[4] Including boats, trains, planes, automobiles, trucks, and ships.

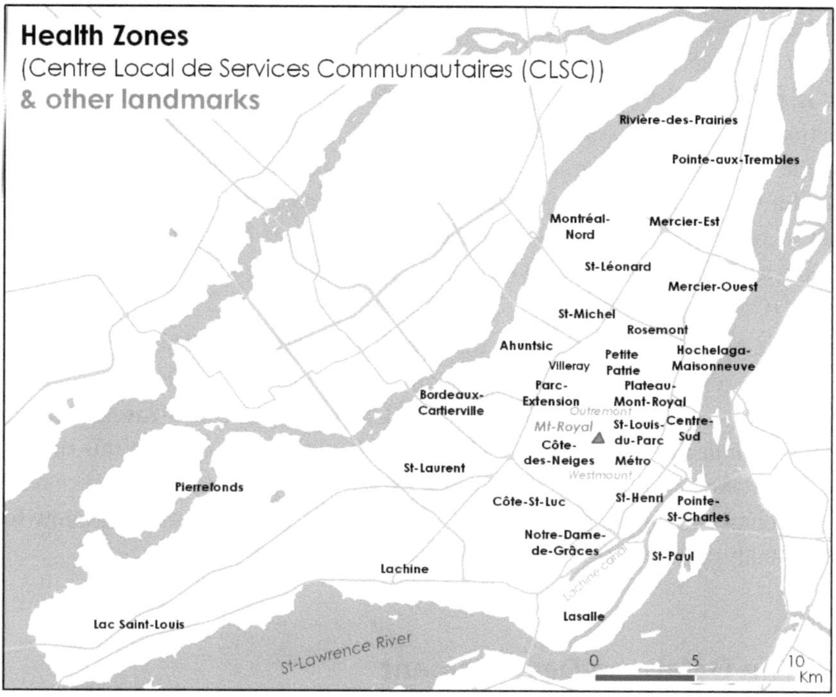

Figure 2.1 The 29 Centre Local de Services Communautaires (CLSC) zones on the island of Montréal.

direct correlation between social or financial deprivation and immigrant status or visible minority status in Montréal (Crouse *et al.*, 2009). Despite the lack of ethnic ghettos, per say, Montréal has been found to be a city with a higher level of income disparity between regions and with observable spatial economic segregation when compared to Canadian norms (Ross et al., 2000; Ross et al., 2002).

Increasing income disparities on the Island of Montréal are echoed by the region's life expectancies, which vary by "more than 13 years between different regions" (Crouse *et al.*, 2009, p 975), but determining what proportion of this disparity is due to local spatial variation of environmental hazards remains to be fully answered. The intricate history and geography of Montréal make it a complicated place to look for patterns of environmental injustice. In a 2009 article Crouse, Ross, and Goldberg compare socio-economic data with measured levels of nitrogen dioxide (NO_2), to gain insight into possible regions and groups that were disproportionately exposed to this common vehicular release toxin that is associated with negative respiratory health impacts. They found that deprivation and ethnicity are not directly related in Montréal in the same way as in many other North American metropolises. The specific neighbourhoods that they noted that had elevated concentrations of NO_2 were: Saint-Laurent, Mont Royal, and Westmount (all affluent), and Parc Extension and Lachine (low income) (Crouse *et al.,* p 979). The conclusions they drew from this set of results somewhat contradict the typical norms of environmental injustice, that is, of higher risk associated with higher deprivation. They found that the central locations of Montréal's educational institutions (UQAM, McGill and Concordia) and the unique set of centrally located affluent neighbourhoods have caused most risks for environmental hazards to cross social boundaries. This underscores the need to reflect upon the region specific social context of a location when interpreting associations between levels of deprivation and levels of environmental burden (Crouse *et al.* 2009).

It is this set of intriguing results that set the stage for the students of the graduate ENVS 663 class (Geographic Information Systems for Environmental Impact Assessment) of winter 2010 at Concordia University to further investigate environmental injustice in the city of Montréal. The attempt to achieve this objective was undertaken by first gathering a wide variety of data on different topics related to environmental injustice. The goal was to compile a database that can allow for a deeper analysis of the situation on the island of Montréal.

2.3 The Database Development

Data and databases are the base of any GIS project. Yet database development is usually the most challenging part of a GIS project. Given the wealth of data available, it is often assumed that all the necessary data required for a specific project exist, in the desired format, "somewhere" and just need to be accessed and compiled. In reality, existing geospatial databases almost never perfectly fit the requirement of a specific situation as will be illustrated in this section. The first pedagogical goal of this project was to expose students to the myth of the perfect database by asking them to compile a relevant geospatial database dedicated to the study of environmental injustice on the island of Montréal. The information that was desired was broken into five groups for collection: socio-demography, health indicators, air-quality, environmental hazards, and noise levels. The goal of each student group was then to develop a dataset on their own topic that would include spatial objects associated to the meaningful criteria. Each of these different datasets was then combined and analyzed in the second phase of the project that allowed for a more in depth study of the environmental injustice situation on the island of Montréal.

2.3.1 Sociodemographic Criteria: The Data is Accurate and Comprehensive

Socio-demographic data are the foundations for any environmental injustice study. In Canada, these data are collected and compiled by Statistics Canada through the census. They are collected every five years and available at a rather fine scale (e.g. dissemination area). Even if the systematic use of administrative limits in social sciences has been criticized because of its artificial representation of the reality (Harris and Hazen, 2006), and its lack of accuracy in studies of environmental injustice (Maantay 2002), it still provides a very good picture of the socio-demographic situation of a given area such as the island of Montréal (Figure 2.2). Students in this group compiled relevant demographic variables, including information on single parents, levels of education, income and amounts of visible minorities in each of Statistic Canada's dissemination areas. These data provide an overview of areas that may be especially vulnerable to environmental injustice, including Saint Michel, Park Extension, Cote des Neiges, Montréal North, and northeastern Saint-Laurent. These students found that generally the east, north, and southwest regions of the Island are more vulnerable than much of the west (Bimrah et al. 2010, 8)[5].

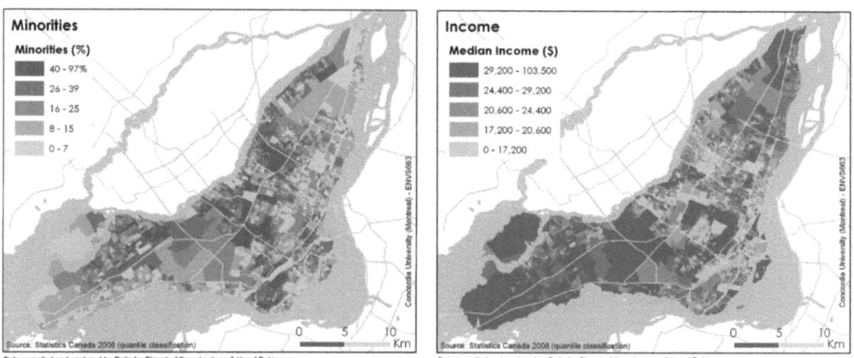

Figure 2.2 Percentage of minorities and average income on the island of Montréal.

2.3.2 Health Information: The Data is Not at the Proper Scale

The second student group focussed on health indicators. The goal of this group was to establish a link between health outcomes and environmental characteristics (Baxter et al. 2010, 4). The major problem with health data is that they are very difficult to access at a fine scale. While the need for health studies at fine scales

[5] For more details on the sociodemographic structure of Montréal, see also the chapter by Tom Weatherburn & Daniel Naud in this volume.

has been demonstrated (Ross et al. 2004), data are often only available at coarser scales due to privacy reasons. In the case of Montréal, relevant health data were only available for the 29 *Centre Local de Services Communautaires*[6] (CLSC), for the 2001 to 2005 period. These data include mortality, total incidences of cancer, lung cancer, respiratory health, incidence of cardiac disease and infant mortality and health (see Figure 2.3).

Students used the Environmentally Attributable Fraction of disease (EAF) (WHO 2006) as well as disease categories particularly indicative of environmental causes (Boyd and Genuis 2008) to define a relative EAF health index[7]. This index was calculated for each CLSC zones, emphasizing three regions with high presence of diseases indicative of environmental causes: Pointe-Saint-Charles, Saint-Henri and Centre Sud (Baxter et al. 2010).

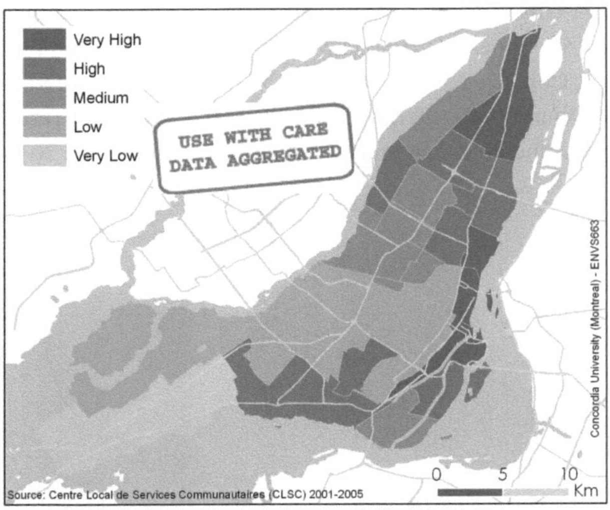

Data compiled and analyzed by AAron Baxter, Charles Cameron, Patrick Culhane, Liohn Sherer

Figure 2.3 Presence of diseases indicative of environmental causes by health zones.

Given the scale discrepancy between the socio-demographic data (fine scale) and the health data (coarse scale), two students decided to aggregate the former to the health sectors (CLSC zones) in order to compare the relationships between socio-demographic profiles and health issues (see Figure 2.4). This comparison was done statistically and visually through the design of a table in which each criterion is classified using a quantile method, which allows the comparison of each health

[6] CLSC's are a form of community health care centre, with a total of 147 in the province of Quebec.
[7] The EAF was developed by the World Health Organization and is a combination of comparative risk assessment data and expert opinions to establish the percentage of mortality and morbidity that are thought to be caused by environmental causes (rather than hereditary).

NUM	CLSC_zone	% CARDIAC	% COPD (Pulmo)	% UNDERWEIGH	% PREMATURE	% RETARDATION	% MORTALITY	% RESPIRATORY MORT	% CANCER	% LUNG_CANCE	Median Income	% Low Income	% single parent	% unempl	% No Diploma	% New_immig
1	Lac-Saint-Louis	0.105	0.267	0.288	0.432	0.270	0.069	0.009	0.058	0.014	93862	9	14	5.9	7.0	3.0
2	Pierrefonds	0.086	0.267	0.390	0.468	0.486	0.083	0.009	0.061	0.017	73335	15	25	6.6	12.0	4.4
3	Dorval-Lachine	0.155	0.479	0.396	0.480	0.504	0.109	0.012	0.068	0.024	60338	22	32	6.4	20.0	4.5
4	LaSalle	0.149	0.463	0.324	0.438	0.504	0.089	0.012	0.072	0.022	55083	24	34	8.3	23.0	5.8
5	Pointe-St-Charles	0.144	0.825	0.474	0.500	0.618	0.133	0.019	0.081	0.043	40514	46	59	12.0	35.0	6.8
6	Verdun	0.139	0.652	0.372	0.474	0.486	0.106	0.017	0.079	0.033	49272	32	47	9.1	23.0	7.7
7	Saint-Henri	0.130	0.538	0.438	0.480	0.564	0.141	0.024	0.072	0.029	45165	42	60	10.0	19.0	5.7
8	Riviere-des-Prairies	0.112	0.385	0.366	0.444	0.498	0.083	0.010	0.065	0.019	63785	18	36	7.2	28.0	2.4
9	Pointe-aux-Trembles	0.123	0.518	0.426	0.492	0.456	0.118	0.016	0.074	0.032	59704	19	31	5.7	26.0	1.2
10	Mercier-Est/ Anjou	0.134	0.534	0.366	0.420	0.540	0.091	0.011	0.070	0.027	57153	23	33	6.5	23.0	4.7
11	Olivier-Guimond	0.136	0.554	0.342	0.474	0.486	0.110	0.015	0.067	0.025	57272	24	33	7.1	24.0	4.5
12	Hochelaga-Maisonneuve	0.111	0.693	0.396	0.450	0.600	0.149	0.022	0.076	0.038	41895	42	51	9.3	27.0	5.5
13	De Rosemont	0.140	0.584	0.366	0.438	0.528	0.088	0.009	0.068	0.025	52877	29	39	8.7	21.0	5.6
14	Saint-Leonard	0.144	0.515	0.336	0.426	0.528	0.071	0.008	0.069	0.021	49352	27	36	10.6	31.0	7.6
15	Saint-Michel	0.127	0.477	0.336	0.420	0.540	0.092	0.010	0.062	0.024	41627	37	47	12.4	38.0	9.1
16	Cote-des-Neiges	0.120	0.230	0.366	0.438	0.570	0.079	0.009	0.056	0.012	51447	34	38	10.2	11.0	13.9
17	Metro	0.114	0.283	0.312	0.390	0.534	0.082	0.012	0.064	0.015	67765	36	38	9.6	5.0	11.6
18	Parc-Extension	0.102	0.249	0.456	0.444	0.604	0.061	0.006	0.052	0.017	34223	53	54	18.1	41.0	18.1
19	Rene-Cassin	0.222	0.310	0.420	0.534	0.516	0.081	0.009	0.063	0.012	65458	24	35	7.4	13.0	8.2
20	NDG-Montreal-Ouest	0.130	0.276	0.318	0.408	0.504	0.084	0.010	0.060	0.018	55667	30	37	9.1	11.0	11.3
21	Saint-Louis-du-Parc	0.080	0.280	0.330	0.408	0.522	0.081	0.008	0.061	0.018	53426	36	46	7.9	15.0	5.5
22	Des Faubourgs	0.117	0.694	0.474	0.552	0.684	0.147	0.018	0.077	0.036	44102	45	53	9.0	19.0	6.5
23	Plateau Mont-Royal	0.079	0.420	0.264	0.366	0.492	0.118	0.013	0.066	0.030	57379	31	32	7.1	11.0	7.0
24	Bordeaux-Cartierville	0.134	0.267	0.360	0.426	0.522	0.080	0.007	0.061	0.014	49776	33	41	11.4	20.0	12.2
25	Saint-Laurent	0.140	0.287	0.354	0.450	0.540	0.071	0.007	0.061	0.016	53018	30	35	9.7	17.0	13.3
26	Villeray	0.137	0.490	0.330	0.408	0.606	0.079	0.009	0.065	0.027	48134	33	42	9.3	24.0	8.0
27	La Petite-Patrie	0.120	0.568	0.312	0.354	0.528	0.106	0.012	0.068	0.029	46269	34	43	8.3	20.0	5.3
28	Montreal-North	0.172	0.586	0.396	0.504	0.522	0.100	0.012	0.068	0.027	42881	38	49	12.5	34.0	7.9

Figure 2.4 *Presence of diseases indicative of environmental causes by health zones. In short, darker means higher risk of environmental injustice for the selected criteria, and vice-versa. From left to right the different columns mean rate (%) of cardiac disease, chronic obstructive pulmonary disease, underweight birth, premature birth, intrauterine growth retardation, mortality, mortality due to respiratory disease, cancer, and lung cancer. The following columns show some sociodemographic characteristics of these health zones (source: Statistics Canada 2006).*

zone across the different criteria. In this table the values have been colour coded based on their quantile rank for each criteria. In short, darker means higher risk of environmental injustice for the selected criteria, and vice-versa. This table emphasizes the existence of CLSC zones with both socio-economic and health indicators of risk of environmental injustice (e.g. 5. Pointe-Saint-Charles), zones with only one set of risks indicators, such as socio-demographic indicators (e.g. 9. Pointe-aux-Trembles), and zones with low risks (e.g. 1. Lac-Saint-Louis).

2.3.3 Air Quality: The Data is Not Available

As discussed previously, on the island of Montréal, air pollution comes mainly from traffic and industrial activities including petrochemical facilities located in the eastern part of the city. Air quality is measured across the island by 17 stations. Unfortunately the data are consistently available only for a dozen of these stations. Given the high spatial level of variability of air quality, there are not enough stations to be able to interpolate air quality on the entire island of Montréal. In order to deal with this problem, students chose two different options to create the air quality database: (1) use an air dispersion model based on traffic; and (2) replicate existing air quality measurements from a previous study.

The dispersion model used to assess the air quality is the California Line Source Dispersion Model (CALINE-4). It is based on the diffusion of gases, derived from the traffic volume and type of traffic. The emission of vehicles was estimated using the U.S. Environmental Protection Agency software Mobile 6.2c. Other inputs were meteorological data, prevalent wind directions and the recorded concentration of NO_2 from the 17 stations of the island. Based on these inputs, the program generates expected concentrations of NO_2 at receptor points. In total, the students created 133 receptor points across the island with NO_2 concentration values in ppm. These values were then interpolated in order to assess air quality variations across the entire island of Montréal (see Fournier et al. 2010 for more details). The main issue with this approach was the lack of traffic data for some sections of roads and highway, as well as the limited number of receptor points due to time constraints.

The second source of air quality data was derived from a the Crouse and colleagues study (2009) already discussed in this chapter, in which researchers had measured the level of NO_2 with Ogawa samplers in 129 dispersed locations on the island of Montréal between 2005 and 2006. This study was intended to provide estimates of intra-urban concentrations of ambient nitrogen dioxide (NO_2) that could subsequently be used in health studies of chronic diseases and long-term exposures to traffic-related air pollution. These data sounded perfectly for this project, but unfortunately it was not made available by the authors of the research. Since the results of this research had already been published, one of the students decided to extract

1001 points from a map appearing in the publication that represented the different ranges of NO_2 concentrations. These points were then interpolated with different methods (mainly IDW and kriging). The results map is one that is very similar to the one generated with the dispersion model (described previously) though with better coverage of the entire island. Although the values of the results are not accurate since they are derived from ranges instead of raw data, the amplitude of the phenomenon is respected and clearly shows the impact of the main highways on the concentration of NO_2 across the island of Montréal (see Figure 2.5).

2.3.4 Environmental Hazards: The Data is Complex

Rather than try to aggregate information on all possible environmental toxins, the environmental hazards group chose to focus more specifically on BTEX (benzene, toluene, ethyl benzene and xylene), a common set of toxins produced by industries,

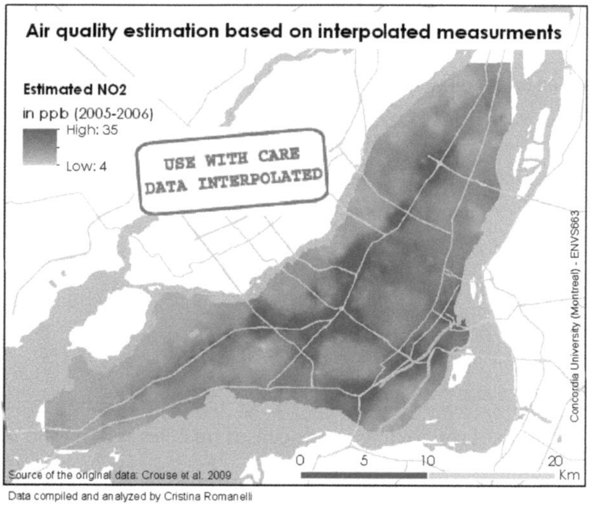

Figure 2.5 General estimates of NO2 based on interpolated data (based on an original map from Crouse et al. 2009).

mainly during petroleum refining[8]. According to the Canadian Council of Ministers of the Environment (CCME 2005), the presence of BTEX - and more specifically of benzene – has been particularly high in Montréal's East End area for decades, due to the presence of two major oil refineries. The main challenge associated with the study of environmental hazards is to assess "the possible implications that the cumulative exposure of simultaneous exposure will have for the environment and for human health" (Marsan et al. 2010, 4). Another challenge presented to this

[8] For a more extensive discussion on this topic see also the chapter by Glenn Brauen in this volume

group was to take into account the different forms of contamination, including air contamination, soil contamination and water contamination. Based on the available data, the students of this group focused on soil and air contaminants.

A list of 541 contaminated sites on the island of Montréal was provided by the Québec *ministère du développement durable, de l'environnement et des parcs*. The 40 chemicals listed for each sites are in terms of presence/absence and there are no concentrations available for individual chemicals. An ordinal BTEX index was created to correspond to the summation of the presence of benzene, toluene, ethyl benzene and xylene for each site. To make these data comparable to other data sets, it has been aggregated at the census level (dissemination area) (see Figure 2.6).

The information about the environmental toxins in the air was obtained from the 2008 National Pollutant Release Inventory (NPRI) which reported industrial sites that released toxins recognized as hazardous to human and environmental health. A weighted BTEX index was then developed to account for the quantitative differences that benzene, toluene, ethylbenzene, and xylene pose to human health following the accepted guidelines established within the United States' Occupational Health and

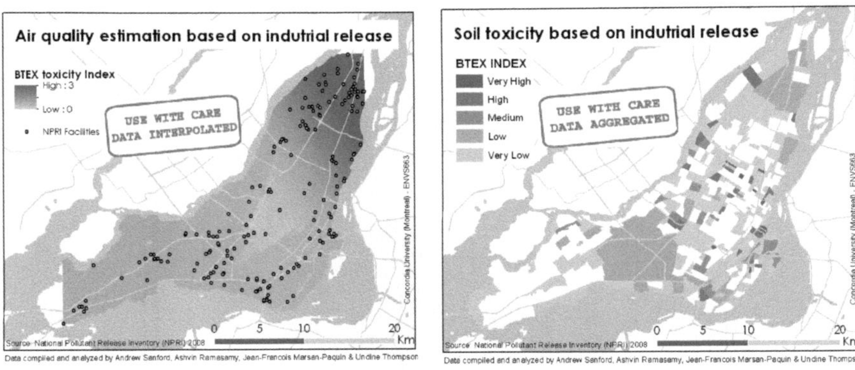

Figure 2.6 Aggregation of soil toxicity at the census track level (left) and interpolation of air quality based on industrial release (right).

Safety Administration (OSHA) standards (NIOSH 2005; Marsan et al. 2010). These data were then interpolated to create a continuous surface of BTEX on the island of Montréal, comparing different interpolation methods (e.g. IDW and kriging) and integrating prominent wind conditions.

The results of this group do not provide an entirely accurate picture of the situation of environmental hazards on the island of Montréal. Nevertheless they emphasize the presence of toxic legacy in soils on many parts of the city, including along the Lachine canal, which was the historical heart of Canadian transportation and industry, as well as the more recent geographical shift of heavy industry toward Montréal's disfavoured north-east (Marsan et al. 2010).

2.3.5 Noise Level: The Data Does Not Exist

Noise is often considered as a major source of dissatisfaction in residential areas, particularly noise originating from infrastructure and industry (Kluijver and Stoter 2002; see also the chapter by Jordan Lacey and Lawrence Harvey in this volume). Noise is also known to also have possible health effects. On the island of Montréal, it is primarily a product of traffic and transportation. The group focussing on noise concentrated on noise originating from transportation infrastructures, namely the airport, highways and commuter train lines. The spatial distribution of noise impact was determined in two steps "(1) Measurement, calculation or prediction of noise level at or near the source; and (2) Predicting noise levels at a distance from the source" (Bobyk et al. 2010, 1). As emphasized by the authors, the impact of noise on communities is affected by its intensity, its duration, the distance form noise source, the existence of barriers, the topography and the atmospheric conditions (Bobyk et al. 2010). These different elements make the modelling of noise in urban environments extremely complex (Farcas and Sivertun 2009). The complexity in this case was aggravated by the paucity of the information.

In order to derive noise level from roads and highways, students used traffic flow data from 1996 to 2008 provided by *Ministère du Transport de Québec* (MTQ 2008). Unfortunately these data were incomplete (60 out of 204 segments had no traffic value). The traffic flow had then to be estimated based on adjacent segments. A Road Traffic Noise Calculator (Java-program (XS4LL 1998)) was used to transform traffic into noise levels in dBA. Finally, a simple noise attenuation model was used to calculate noise propagation up to a distance of 210 m from the source (Canter, 1996). Train traffic was estimated using *Agence Métropolitaine de Transport de Montréal* (AMT) commuter line schedules and a formula based on the average number of trains per hour and the sound emitted by a passing train (Goff and Novak 1977).[9] The model developed previously for roads and highways was run again to produce a standardized map of railroad noise for a distance of 240 meters. Finally, the airport noise levels were provided by *Aéroports de Montréal.* Unfortunately, these data were not available in dBA, but in Noise Exposure Forecast (NEF). The combination of these different noise sources required their conversion into a common system based on impact significance (e.g. acceptable, low disturbance, unacceptable). A map of transportation noise impact on the island of Montréal derived from transportation sources was created. Given the many issues associated with the poor quality of the data available in this topic (e.g. lack of data, lack of parameters for running the noise models, different unit systems), it was decided to not incorporate it in the final database.

[9] Note: The data for freight trains were not available.

2.4 Discussion

Through this process, students were able to compile collectively a geographic database. Although this database had some obvious limitations (see Table 2.1), its development served three major goals: (1) Expose students to the complexity and flaws of database development; (2) Investigating the general trends in terms of environmental injustice in Montréal; and (3) Provide a point of departure for the exploration of the idea of mapping a database which was the topic of the workshop organized in Montréal in September 2010.

Table 2.1. Synthesis of the sources and quality of the data compiled in the database.

		Geography		Attributes		Overall	Source
		Scale	Unit	Content	Completeness		
Socio-demographic		Fine	DA	Rich	Complete	Good	1
Health		Coarse	CLSC zones	Rich	Complete	Fair	2
Air quality	Model	Fine	Source points	N/A	Incomplete	Poor	3
	NO2	Fair	Surface map	Fair	Complete	Fair	4
Hazard	Soil	Fine	Source points	Fair	Complete	Fair	5
	Air	Fine	Source points	Rich	Complete	Fair	6
Noise		Coarse	Line + isolines	Poor	Incomplete	Poor	7

1) Statistics Canada (2006); 2) Carrefour Montréal, Atlas Santé (2001-2005); 3) MTQ (1996-2008) + Ville de Montréal; 4) Crouse et al. (2009); 5) Ministère du développement durable; 6) NPRI (2008); 7) MTQ (1996-2008) + Aéroports de Montréal.

The process of compiling the database was clearly pedagogical. Throughout this process, students were exposed to the major issues related to database development such as: (a) the data have never been collected therefore they do not exist: Can they be assessed? (e.g. noise data across the island); (b) the data exist but they are not available for confidentiality reasons (e.g. Health data) or for other types of reasons (e.g. NO_2 data); (c) the data exist but is incomplete (e.g. road traffic); (d) the data exist, but not in the right format or at the proper scale (e.g. Health data for CLSC zones); (e) the data exist but not for the appropriate time period. It is also important to mention that a comprehensive database to study environmental injustice would include other types of data such as access to vegetation (see chapter by Hiên Pham and colleagues in this volume), pedestrian accidents, drinking water quality, and access to healthy food. This database is far from being comprehensive and perfectly accurate, but throughout the process of compiling it, students were re-examining their own notion of database comprehensiveness and accuracy in studying complex issues such as environmental injustices.

Beyond these limitations, the database was rich enough to be analyzed in more detail. This was the goal of the second part of this course project. Through GIS analysis and critical analysis of portions of the amalgamated data, students attempted

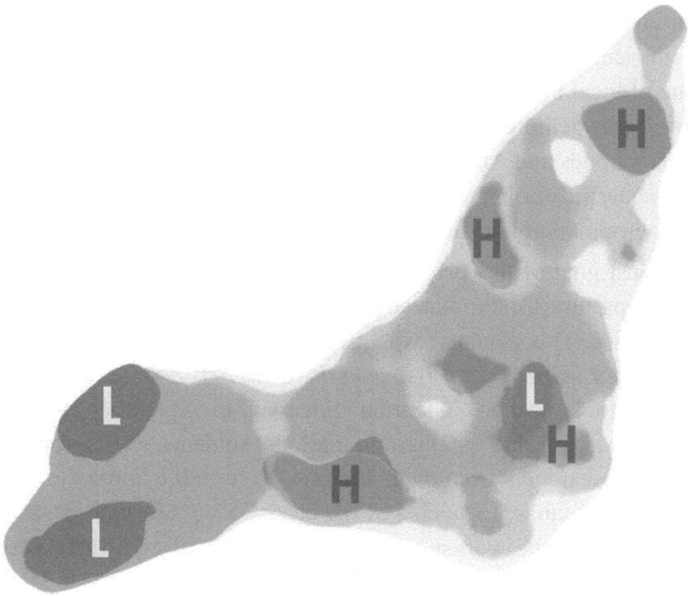

Figure 2.7 Attempt to visually mitigate the "overly precise" results due to the data limitation as well as to the limits of the methodology used to analyze these data.

Green zones (L) show areas of Low Risks of Environmental Injustice and the red zones (R), show areas of high risks (Author: Aaron Baxter. Used with permission).

to shed light onto the topic of environmental injustice in Montréal. Geospatial methods such as spatial interpolation (inverse distance weighting, kriging), and spatial statistics (e.g. clustering analysis) were used to investigate if any meaningful patterns would emerge from the data. Given the issues associated to the quality of the original data and the difficulty of comparing such a wide range of information, it was beyond the scope of this second part of the project to try to provide sound final results as is illustrated in Figure 2.7. Nevertheless, these different analyses allow sketching some general trends and pointing to some areas. A few of Montréal's neighbourhoods recurrently appear as regions that exhibit higher than average risk for environmental injustice. These neighbourhoods include, Pointe-Saint-Charles, Saint-Henri, Mercier-Hochelaga-Maisonneuve, as well as Parc Extension. These are areas that should be more carefully examined in order to gain a clearer understanding of the complicated dynamics of environmental injustice on the Island of Montréal.

The results also show the specificity of the situation on the island of Montréal. While criteria to study environmental injustice come from the United States where racial segregation is much more prevalent than in Montréal, the percentage of visible minorities did not appear has being a key element in defining areas at risk of environmental injustice, as also emphasized by Crouse, Ross, and Goldberg (2009). The possible issue of local Francophones living in the more environmentally hazardous areas of Montréal is definitely worthy of note, as it may reflect historical struggles between Francophones and Anglophones. This should certainly be expanded upon and critically analyzed further, as it may be able to help expand the larger understanding of environmental injustice and its underlying causes.

Finally, the last goal of this project was to provide a database that could serve as a point of departure to stimulate thought and discussion around the issue of mapping a database. This database was given to the participants of the workshop entitled *'Mapping' Environmental Issues in the City: Arts and Cartography Cross-Perspectives* that took place at Concordia University in September 2010. The goal of this workshop was to bring together international students, artists, cartographers, geographers and researchers from the humanities to further explore the meaning of mapping a specific database. Students of the class ENVS 663 had paved the path, and the participants in the workshop took it in many different directions as this volume illustrates.

2.5 Conclusion

This chapter has expanded upon the topic of environmental injustice as a way of approaching the uneven burden of health implications of negative externalities from development in general, and within Canada and Montréal more specifically. It has outlined the history of environmental injustice theory within the United States and its development in Canada and Montréal. The issue of environmental injustice is not straight forward in Montréal, as its historic and socio-economic context are unique in North America. The outcomes of the project presented in this chapter seem to point that there are areas of the city that are at heightened risk for health implications due to environmental "bads." What appears more clearly throughout this project is the possibility to compile a decent database by pulling together existing resources, as well as the simultaneous impossibility to develop a comprehensive and accurate database without producing our own data. As pointed out by Denis Wood (2010), the cost of compiling databases about the environment is insane, which explains why we have to rely so heavily on existing databases provided by institutions and governments to study the environment. Some of these data – at least in Canada - provide an accurate and comprehensive basis to study environmental issues (e.g. sociodemographic data), but most of these data are usually not available at the proper scale (e.g. Health data), or for the entire zone (e.g. Air quality and traffic), or with the relevant attributes (e.g. Environmental hazards), or for the appropriate period of time. Throughout this process the utopian goal of compiling a comprehensive and reliable database to study environmental issues unveiled slowly, as well as the many limitations of using GIS to address such a complex issue.

At the end of this chapter it is important to emphasize that sound studies of environmental injustice require much more than simply accurate data and sophisticated GIS analyses. It requires an in depth understanding of the historical, social, demographic and geographic situation both at the very local scale and at the transnational level as well. It requires inputs from many different domains of the arts, sciences and the humanities. This project was only a small step in the direction of exposing students to the complexity of developing a GIS database, which was already a respectable pedagogical achievement.

Acknowledgements

We would like to thank the following students of the winter 2010 graduate course entitled "GIS for Environmental Impact Assessment" (ENVS 663) at Concordia University who compiled the database: Aaron Baxter, Rajinder Bimrah, Roksolana Bobyk, Charles Cameron, Patrick Culhane, Bastien Fournier-Peyresblanq, Hans Ghoorbin, Iffat Huque, Vassilakis Konstadinos, Allison Lapierre, Jean-Francois Marsan-Paquin, Anita Ogaa, Ashvin Ramasamy, Yosef Robinson, Cristina Romanelli, Andrew Sanford, Liohn Sherer, & Patrick Tanyi.

References

Baxter A., Cameron C., Culhane C., and Sherer L. (2010). *Health data for assessing environmental injustice on the island of Montréal*, Student Report, Concordia University, department of Geography, Environment and Planning.

Bimrah R., Robinson Y., and Lapierre A. (2010). *Socioeconomic Term Project Phase 1*, Student Report, Concordia University, department of Geography, Environment and Planning.

Boyd, D. R., & Genuis, S. J. (2008). The environmental burden of disease in Canada: Respiratory disease, cardiovascular disease, cancer, and congenital affliction. *Environmental Research*, 240-249.

Brody, H. (1997). *Maps and Dreams.* Vancouver: Douglas & McIntyre.

Bullard, R. (1990). *Dumping in Dixie: Race, class, and environmental quality.* San Fransisco: Westview Press.

Buzzelli, M. (2008). *Environmental justice in Canada - It matters where you live.* Ottawa: Canada Policy Research Networks (CPRN).

Buzzelli, M., & Jerrett, M. (2007). Geographies of susceptibility and exposure in the city: Environmental inequity of traffic-related air pollution in Toronto. *Environmental Hazards, 30* (2), 195-210.

Buzzelli, M., Jerrett, R., Burnett, & Finke, N. (2003). Spatiotemporal perspectives on air pollution and environmental justice in Hamilton, Canada. *Annals of the Association of American Geographers, 93*, 557-573.

Canter L. (1996). "Prediction and Assessment of Impacts on the Noise Environment". *Environmental Impact Assessment*. McGraw-Hill: New York

Crouse, D. L., Ross, N. A., & Goldberg, M. S. (2009). Double burden of deprivation and high concentrations of ambient air pollution at the neighbourhood scale in Montréal, Canada. *Social Science & Medicine, 69*, 971-981.

ENVS 663. (2010). Environmental Justice Results. Retrieved May 2011, from https://sites.google.com/site/envs663/2010

Farcas F, and Sivertun A. (2009). Road Traffic Noise: GIS Tools for Noise Mapping and a Case Study for Skane Region, *The International Archives of Photogrammetry, Remote Sensing and Spatial Information Sciences, 34*, Part XXX.

Fournier B., Ghoorbin H., Romanelli C., and Tanyi P. (2010). *Mapping Air Quality on the Island of Montréal*, Student Report, Concordia University, Department of Geography, Environment and Planning.

Goff, R.J. and Novak, E.W. (1977). Environmental Noise Impact Analysis for Army Military Activities: User Manual, *Tech. Rep. N-30, U.S. Army Construction Engineering Research Laboratory*.

Gower, S., Hicks, J., Shortreed, J., Craig, L., & McColl, S. (2008). Development of a Health Effects-Based Priority Ranking System for Air Emissions Reductions From Oil Refineries in Canada. *Journal of Toxicology and Environmental Health, Part A, 71*, 81–85.

Harris, L. & Hazen, H. D. (2006). Power of Maps: (Counter)-mapping for Conservation. *ACME International E-journal of Critical Geographies, 4(1)*, 99-130.

Langlois, A., & Kitchen, P. (2001). Identifying and measuring dimensions of urban deprivation in Montréal: an analysis of the 1996 census data. *Urban Studies, 38* (1), 119-139.

Ley, D. F. (1993). *Past elites and present gentry: neighbourhoods of privilege and the inner city.* Montréal: McGill-Queen's University Press.

Lobmayer, P., & Wilkinson, R. (2002). Inequality, residential segregation by income, and mortality in U.S. cities. *Journal of Epidemiology and Community Health, 56* (3), 183-187.

Marsan, J.F., Ramasamy, A., Sanford, A., Thompson, U. (2010). *Environmental Hazards on the Island of Montréal: Toxic Pollutants.* Student Report, Concordia University, Department of Geography, Environment and Planning

Montréal. (2010). *Encyclopaedia Britannica.* Retrieved May 10th, 2011, from http://0-search.eb.com.mercury.concordia.ca/eb/article-12462

Robbins, P. (2005). *Political ecology: A critical introduction.* Oxford: Blackwell.

Romanelli, C. (2010). *Arts & cartography workshop: Mapping environmental issues in the city*. Retrieved May 14th, 2010, from database: http://mappingworkshop.wordpress.com/database/

Ross, N. A., Tremblay, S., Graham, K. (2004). Neighbourhood influences on health in Montréal, Canada. *Social Science and Medicine*, 59, 1485 - 1494

Ross, N. A. (2004). *What have we learned studying income inequality and population health?* Ottawa: Canadian Institute for Health Information.

Ross, N. A., Nobrega, K., & Dunn, J. (2002). Incom segregation, income inequality and mortality in North American metropolitan areas. *Geojournal, 53* (2), 117-124.

Reseau De Surveillance De La Qualite De L'Air. (2007). Air Quality in Montréal, RSQA Annual Report. http://ville.Montréal.qc.ca/portal/page?_pageid=4537,7190968&_

dad=portal&_schema=PORTAL (retrieved December 2010).

Statistics Canada. (2006). 2006 census data.

Townsend, P. (1987). Deprivation. *Journal of Social Policy, 16* (2), 125-146.

U.S. E.P.A. (2010). *Environmental Justice*. Retrieved May 10th, 2011, from United States Environmental Protection Agency: http://www.epa.gov/environmentaljustice/

Westra, L. (2008). *Environmental justice and the rights of indigenous peoples.* London: Earthscan.

White, H. (1998). Race, class, and environmental hazards. In D. Camacho (Ed.), *Environmental injustices, political struggles.* Durham, NC: Duke University Press.

WHO (World Health Organization). 2006. *Preventing Disease through Healthy Environments*. WHO Press: Geneva, Switzerland.

Wood, D. (2010). *Rethinking the Power of Maps*. New-York: The Guilford Press.

Chapter 3

Observations

Mapping the Greenscape and Environmental Equity in Montreal: An Application of Remote Sensing and GIS

Thi-Thanh-Hiên Pham[1], Philippe Apparicio[1], Anne-Marie Séguin[1], and Martin Gagnon[1-2]

[1] Institut national de la recherche scientifique - Urbanisation Culture Société, Montréal, Québec.
[2] Institut d'urbanisme, Université de Montréal, Montréal, Québec.

Abstract

Vegetation and green spaces provide multiple benefits for urban life but they are not always evenly distributed throughout cities. Recent studies have shown that deprived and ethnic populations have less access to vegetation, which is a form of environmental inequity. The goal of this study is to map the vegetation cover and to spatially depict the problematic areas in terms of environmental inequity in Montreal. We carry out an object-oriented classification in eCognition from Quickbird images (at a resolution of 60cm) to identify two categories of vegetation: trees/shrub and grass. We then compute 12 vegetation indicators representing the proportion of vegetation, trees/shrub and grass in streets, alleys and residential yards. Finally, statistical analyses are undertaken to reveal the link between the vegetation indicators and the proportion of immigrants, visible minorities and low income individuals. Our results show that the proportion of vegetation varies significantly across the boroughs. About 30% of the areas exhibiting an elevated proportion of the three groups are identified as very high inequity whereas 10 to 14% are identified as areas with high green benefits. Environmental inequity with respect to the three populations also expresses differently depending on the type of green spaces (street and alleys for immigrants and visible minorities while all the three types of green spaces for the low income population). This study may interest city planners and local governments as the findings could better inform decisions regarding greening programs.

3.1 Introduction

Increasing concerns about sustainable cities are bringing urban green spaces to the forefront of both academic and governmental discourses, notably in the context of global warming. It is well established that green spaces play a vital role in the quality of life of urban dwellers. Green spaces provide cooling effects, reduce air and noise pollution, fix carbon and filter storm water runoff and thus contribute to energy savings (Jo and McPherson 2001; Ridder et al. 2004; Witford et al. 2001; McPherson et al. 2005). In terms of social benefits, green spaces help reduce stress levels and promote social integration of older adults and children, especially in a multi-ethnic context (Cackowski and Nasar 2003; Kweon et al. 1998; Castonguay and Jutras 2008; Seeland et al. 2009). However, urban green spaces are threatened by real estate development. According to the *Conseil régional de l'environnement de Montréal*, between 1998 and 2005, the vegetation cover in the Montreal Metropolitan Community diminished by 18%, equivalent to 310 ha (CREMTL 2008). Moreover, another study provides evidence that green spaces are not equally distributed within Canadian cities and poor neighbourhoods tend to have less vegetation than affluent ones (Tooke et al. 2010), suggesting the existence of environmental inequity (Walker 2009). A recent report of Canadian Institute for Health Information (CIHI 2011) also shows that areas with low socioeconomic status are 10 times more likely to be exposed to elevated land surface temperature than affluent areas.

To protect existing green spaces, create new ones and tackle current inequalities in their distribution in highly complex urban settings, there is an urgent need to inventory the vegetation cover at a very fine resolution. In turn we will be able to identify areas exhibiting deficient vegetation and a high proportion of a particular population group. Consequently, the objective of this study is twofold. First, we aim to map vegetation forms (tree, shrubs, grass) by using very-high-resolution images and calculate the proportion of each vegetation form in different elements of urban fabric (streets, alleys, back- and front-yards) in Montreal. Second, we investigate the extent of environmental equity related to green spaces by mapping problematic areas with regard to potentially vulnerable populations, such as the underprivileged, immigrants and visible minorities.

3.1.1 Green Spaces Mapping by Using Remote Sensing and GIS

Three approaches are commonly used to produce a detailed vegetation map from remote sensing data: visual interpretation, pixel-based classification and object-based classification. The first one is based on human interpretation and manual delimitation. Despite its high accuracy, this approach is subjective, time-consuming and costly (Freeman and Buck 2003; Akbari et al. 2003). The second group of methods rely on pixel's spectral responses. Conventional pixel-based methods using solely on pixel's spectral responses are claimed to be less precise to extract urban objects as the inter-class spectral confusion and intra-class spectral variation increases significantly in urban settings. Several advanced methods are being developed in order to resolve this problems, namely kernel-based reclassification, fractal techniques or spectral unmixing (Kobler et al. 2006; Myint and Lam 2005; Tooke et al. 2010). The third group of methods, object-based, offers another alternative which incorporate spectral, spatial and context information into the classification.

The debate on either pixel-based or object-based approach gives the best accuracy still continues and it is likely that both of them have the advantages and disadvantages depending on the study area, the classes to be extracted as well as the processing time (Thomas et al. 2003; Mallinis et al. 2008; Bock et al. 2005). However, in a few tested cases, an object-based method outperforms pixel-based conventional method (see for example Myint et al. (2011) and Yu et al. (2006)). Compared to advanced methods per pixel, the object-based approach is also easy to implement, allows integrating auxiliary data and produces classifications in vector format which can then be integrated into a GIS database (Yu et al. 2006). For these reasons we resort to the later approach to extract the vegetation cover in urban green spaces in Montreal.

A growing number of researchers are using the object-based approach to obtain different types of urban vegetation from aerial photography or satellite images. Most of them manage to identify three types of vegetation, namely trees, grass and shrubs (Damm et al. 2005; Zhang et al. 2010; Mathieu et al. 2007). Accuracy is good enough for the trees and grass classes, i.e. 60% to 90% or even higher in Zhang et al. (2010) thanks to the combination of several image analysis algorithms. However, accuracy is always relatively low for the shrub class (30-40%). To improve accuracy, two solutions are commonly used: manually editing objects by experts who are familiar with the study area (Mathieu et al. 2007) or combining optical images with height information extracted from LiDAR data or GPS measurements (Zhou and Troy 2008; Iovan et al. 2007; Delm and Gulinck 2009). However, LiDAR data are costly and not always available (such is the case of our study area).

In the light of this literature review, we attempt to isolate three classes of vegetation from very-high-resolution images –trees, shrubs and grass–, which will be combined with land-use information to map the greenscape in Montreal.

3.1.2 Environmental Equity Related to the Vegetation Cover

There is increasing evidence that poor neighborhoods characterized by low incomes and the presence of racial or ethnic minorities have less vegetation than wealthier neighborhoods. For example, in Chicago, Iverson and Cook (2000) observed a positive correlation between household income and the presence of public and private trees. Landry and Chakroborty (2009) noted that there are fewer street-side trees in neighbourhoods of Tampa inhabited by black communities, low-income people and renters. A similar relationship was documented by Heynen et al. (2006) for the Hispanic communities in Milwaukee. In contrast, Troy et al. (2007) note that in Baltimore, even if the home value has a positive impact on the presence of trees, African-Americans are overrepresented in areas where there are a lot of trees.

In Montreal, Tooke et al. (2010) demonstrate that the presence of immigrants shows no significant relationship with the amount of vegetation. They also note a negative correlation between the amount of vegetation and the percentage of low income individuals. However, after controlling for education and immigrant status by computing a geographically weighted regression model (GWR), they found only three census tracts exhibiting low vegetation and low median family income. This study is highly interesting because it compares the trends in the three major Canadian cities. Yet some local disparities may remain hidden because of the geographical scale and the lack of consideration for how the vegetation is used and managed (for example, in public parks, private gardens or vacant lots).

3.1.3 Hypothesis

In this study, we observe the green spaces more carefully to determine the extent of environmental equity related to different types of green spaces in Montreal. Based on our knowledge of the study area, we think, contrary to Tooke (2010), that it is relevant to verify not only the relationships between vegetation and the low income population, but also with the immigrant and visible minority populations. We put forward the hypothesis that low income individuals, immigrants and visible minorities will be overrepresented in areas where the vegetation is less abundant.

3.2 Materials and Methods

3.2.1 Study area

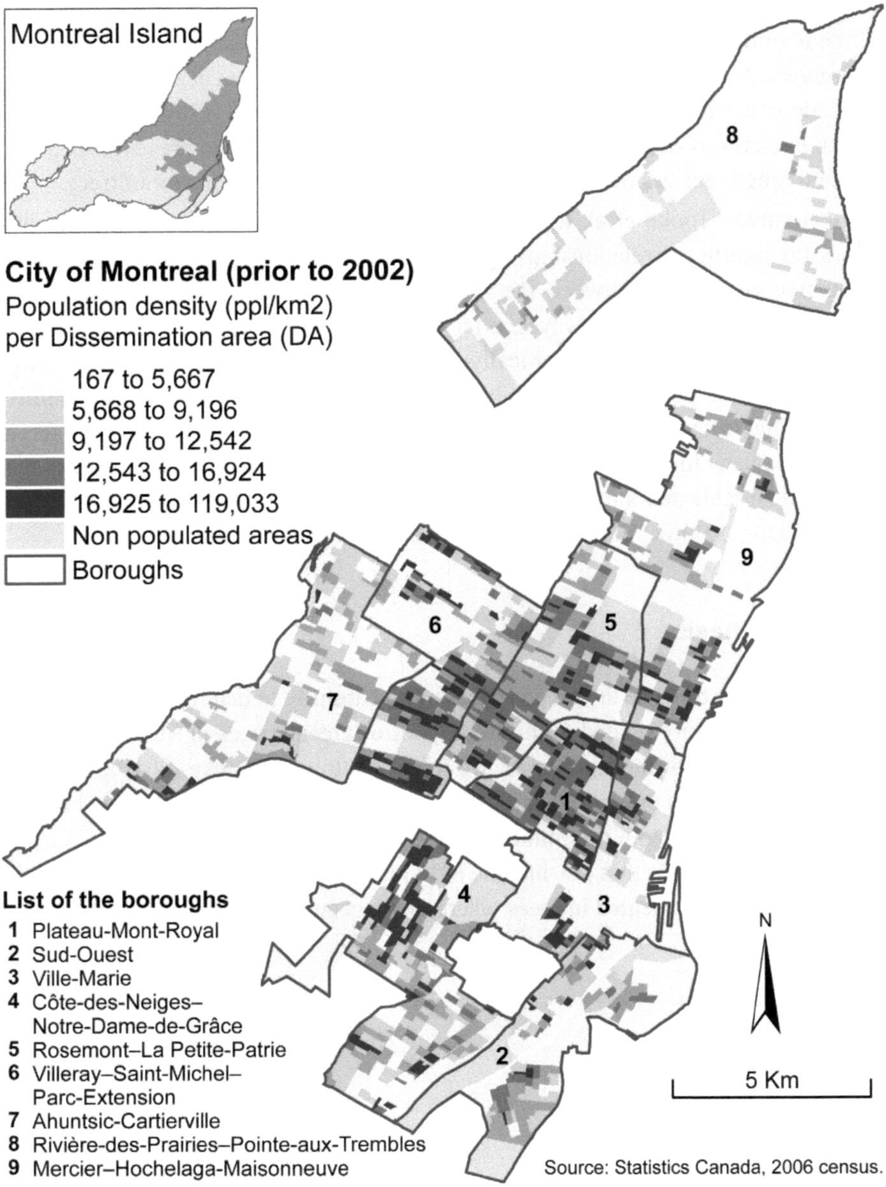

Figure 3.1 Study area.

The study area corresponds to the territory of the City of Montreal before the municipal mergers of 2002. This choice was made based on the fact that the cadastral dataset was only available for this area. It is composed of nine boroughs covering 184 km^2 (Figure 3.1). In 2006, there were 1,033 302 people living in this area. The population density varied from 120,000 inhabitants/km^2 in very dense areas (such as Le Plateau Mont-Royal) to 5,000 inhabitant/km^2 in inner-ring suburbs (Rivière-des-Prairies–Pointe-aux-Trembles). The city is characterized by more than 700 urban parks including numerous pocket parks, sport parks and regional parks. Other types of green spaces that are common here include streets, alleys and gardens.

3.2.2 Remote Sensing and GIS Data

The vegetation cover is extracted from two Quickbird images acquired in September 2007. Their four multispectral bands and panchromatic band are fused with the Pansharp tool of the PCI software application (Zhang 2002). We conduct an ortho-rectification on the images with the rational model of PCI (Toutin et al. 2002) by using 23 GCP points (XRSM and YRSM errors are respectively 1.83m and 1.37m).

The satellite images are complemented with a building footprint dataset produced in 2003 (provided by the City of Montreal). Once obtained, the vegetation classes are juxtaposed with maps indicating three land-use types: streets, alleys, and back- and front-yards. These data are extracted from block and parcel maps (also provided by the city).

3.2.3 Vegetation Classification

The object-based classification is carried out in eCognition 8.1 software. First of all, we create homogenous objects with an algorithm minimizing the overall heterogeneity of objects (segmentation step). The algorithm is a function of several parameters which are defined by users: bands, the scale factor, colour/shape ratio and compactness/smoothness ratio (for further details see the eCognition user manual).

We carry out the segmentations on the four original bands, a building map and a Modified Soil Adjusted Vegetation Index (MSAVI) band recommended by Delm and Gulink (2009). The MSAVI band is calculated with the following equation (Qi et al. 1994):

$$MSAVI2 = \frac{2\rho NIR + 1 - \sqrt{(2\rho NIR + 1)^2 - 8(\rho NIR - \rho RED)}}{2} \quad (1)$$

where ρNIR and ρRED are the reflectance of the near infrared band, red band and green band, respectively.

The scale factor represents the size of the objects to be obtained. We segment the images at two levels (50 and 15) and at each level we use the same parameters of colour and compactness. The colour/shape ratio is chosen to be 0.9/0.1 as the colour of the vegetation is different from other categories of land use. The compactness/ smoothness ratio is set as 0/1 allowing us to obtain objects with low compactness which tend to correspond to tree groups as opposed to objects with high compact-

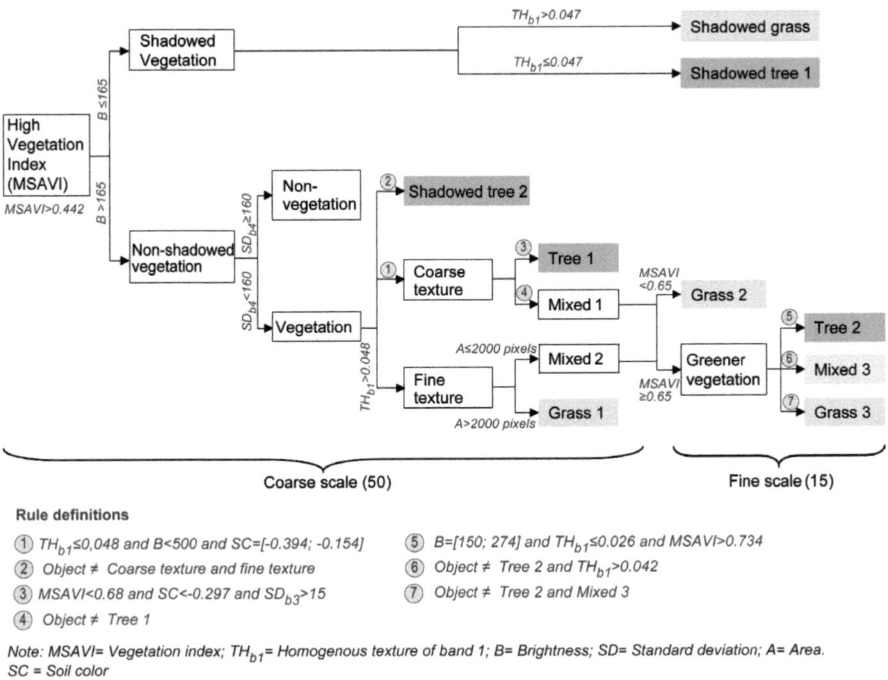

Figure 3.2 Image processing steps conducted in eCognition and classification rules.

ness which tend to correspond to buildings and man-made features.

Once we achieve a satisfactory segmentation, we classify objects into classes using one or several rules composed of spectral, textural and class-related information (see also Figure 3.2). Note that Soil Color (Mathieu et al. 1998) is computed as follows:

$$SoilColour = \frac{\rho RED - \rho GREEN}{\rho RED + \rho GREEN} \qquad (2)$$

where ρRED and ρGREEN are the reflectance of the red band and green band respectively.

Finally, to evaluate the accuracy of our method, we carry out a manual delineation of three types of vegetation on 50 sites from the Quickbird images: tree groups, grass surface and mixed. The sites (200mx200m) are distributed randomly on the city totalizing about 1% of the study area, as recommended by Congalton and Green (2009). The delineation is used as a reference for computing the confusion matrix for three classes. However, as the *Mixed* class shows important confusions with the *Tree* class, we regroup them into one single class named "tree/shrub" in the final classification.

3.2.4 Environmental Equity Analysis

The vegetation data extracted from the images and city's maps are aggregated into the dissemination area (DA) defined by Statistics Canada. A DA is composed of one or a few blocks with a population ranging from 400 to 700 people. Our statistical analyses are then carried out at the DA level. We obtain 12 vegetation indicators: the proportion of vegetation and its forms (tree/shrub, grass) by DA and the proportion of vegetation and its forms in streets, alleys or yards by DA. To measure environmental equity, three sociodemographic variables are selected, based on the literature review: the percentage of immigrants, the percentage of visible minorities and the percentage of low income people (see statistics of the variables in Table 3.1).

Table 3.1. Summary of census variables for the City of Montreal.

	Mean	Std Dev	Minimum	Maximum
Immigrant (%)	29.33	17.95	0.00	79.31
Visible minorities (%)	24.99	20.45	0.00	96.67
Low income (%)	32.90	16.14	0.00	94.40

(Number of dissemination areas: 1,773)

In order to investigate the existence of environmental equity, we first compute Pearson correlation coefficients between the vegetation indicators and the sociodemographic variables.

Next, we map the problematic areas exhibiting different levels of environmental equity (as illustrated in Figure 3.3). The first step is to select the DAs with a high proportion of a given population group, i.e. 40% of the DAs with the highest values (P60 to max). Then we cross-tabulate this selection with the quintiles of each of the vegetation indicators. Thus, we can identify problematic DAs which have both a high proportion of a given group and low values of a given vegetation indicator: **very high inequity** (DAs belonging to the first quintile of the vegetation indicator: Min to P20) and **high inequity** (DAs belonging to the second quintile: P20 to P40). In the same way, non-problematic areas include DAs having a value of vegetation indicator belonging to the third quintile (**no disadvantage and no benefit**: P40 to P60), and DAs having a higher proportion of vegetation (**green benefits**: P60 to P80; **high green benefits**: P80 to max).

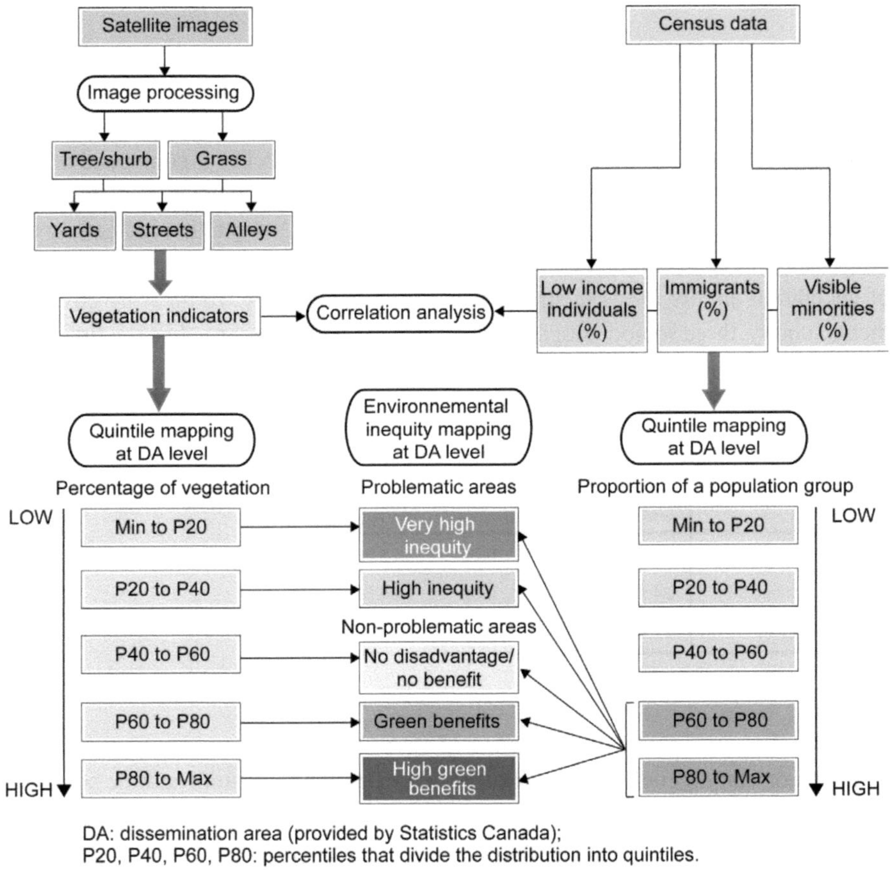

Figure 3.3 Methodological diagram.

3.3 Results

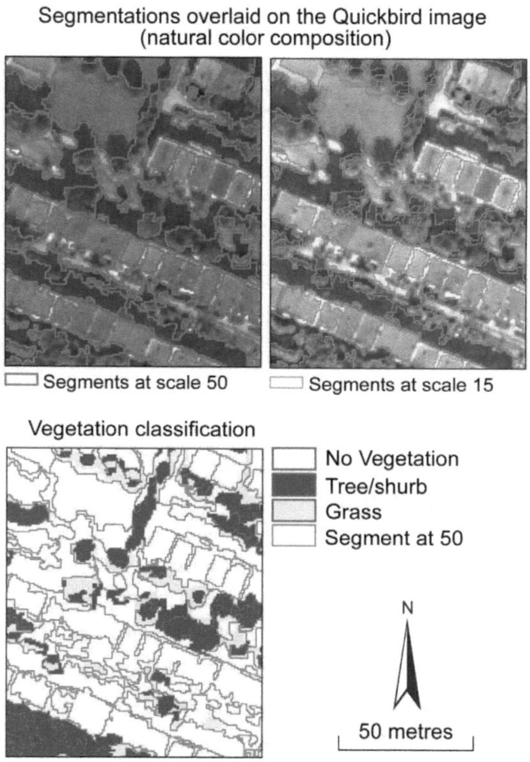

Figure 3.4 Examples of segmentations and classification by the object-based method.

3.3.1 Vegetation Classification

Examples of segmentations and classification of the object-based method are given in Figure 3.4. This method produces a relatively precise classification. In fact the overall accuracy is about 74%, the producer's accuracy of *Tree/shrub* and *Grass* classes are 77.5% and 70.7% (Table 3.2). These accuracy rates are close to those obtained by previous authors. Damm et al. (2005) obtained 64% and 84% for their class of "tree-shrub" and "herbaceous plants"; Mathieu et al. (2007) also obtained 63% and 70% for the class "tree" and 66% and 90% for the class "grass". In most of the cases in Montreal, the grass surface is well maintained and hence appears on the QuickBird images with a fine texture that is easily identifiable. However, there are also commission errors (19.33%) which occur where grass and bare soil are interspersed or where grass is masked by the shade of trees and buildings. As for *Tree/shrub*, we note two main sources of errors: *Tree/shrub* masked by buildings' shadows or *Tree/shrub* confused with grass in vacant lots.

Table 3.2. Confusion Matrix

		Classification			
		Tree/shrub (m^2)	Grass (m^2)	Total (m^2)	User's acc.[b] (%)
Reference	Tree/shrub (m^2)	**281,596**	141,276	422,872	66.59
	Grass (m^2)	81,8690	**341,613**	423,483	80.67
	Total (m^2)	363,465	482,889	**846,355**	
	Prod.accuracy[a] (%)	77.48	70.74		73.63

[a] This rate refers to the probability of a reference pixel being correctly classified.
[b] This rate refers to the probability that a pixel classified on the map represents that cate-gory on the ground (Congalton and Green 2009).

3.3.2 Greenscape Evaluation

Once the images are classified and juxtaposed with the DA map, it is now possible to evaluate the presence of vegetation across the city's boroughs (Table 3.3 and Figure 3.5a to c). Overall, 26.73% of the city's area is covered by vegetation, i.e, nearly 10.5% of the territory being covered by trees and shrubs whereas 16% by grass. The ANOVA analysis shows that the amount of vegetation varies significantly throughout the city (R^2=0.32). For most of the boroughs, the proportion of grass is more important than that of trees and shrubs, except Côte-des-Neiges—Notre-Dame-de-Grâce.

Table 3.3. Vegetation, tree/shrub and grass cover by boroughs.

Boroughs	Total vegetation in the DA (%)	Tree/shrub in the DA (%)	Grass in the DA (%)	Total Veg. in streets by DA (%)	Total Veg. in alleys by DA (%)	Total Veg. in yards by DA (%)
1. PMR	16.74	6.75	9.98	18.15	23.99	18.39
2. SO	26.10	9.29	16.80	18.76	19.28	30.60
3. VM	16.30	6.03	10.27	10.78	14.05	14.62
4. CNNDG	34.10	17.65	16.45	25.73	12.10	42.99
5. RPP	24.08	9.84	14.24	24.05	20.40	30.32
6. VSMPE	18.26	7.26	11.00	13.99	14.83	26.68
7. AC	37.88	18.04	19.84	22.39	18.34	47.44
8. RDPPAT	38.17	10.72	27.45	13.81	5.25	41.44
9. MHM	28.97	9.45	19.52	21.32	20.54	34.86

See Figure 3.1 for borough names.

Boroughs	Total vegetation in the DA (%)	Tree/shrub in the DA (%)	Grass in the DA (%)	Total Veg. in streets by DA (%)	Total Veg. in alleys by DA (%)	Total Veg. in yards by DA (%)
8. RDPPAT	38.17	10.72	27.45	13.81	5.25	41.44
City	26.73	10.56	16.17	18.78	16.53	31.93
ANOVA R^2	0.3220	0.3441	0.3301	0.1430	0.1688	0.2741
ANOVA Fisher-F	104.70	115.70	108.66	36.78	28.57	83.27

See Figure 3.1 for borough names.

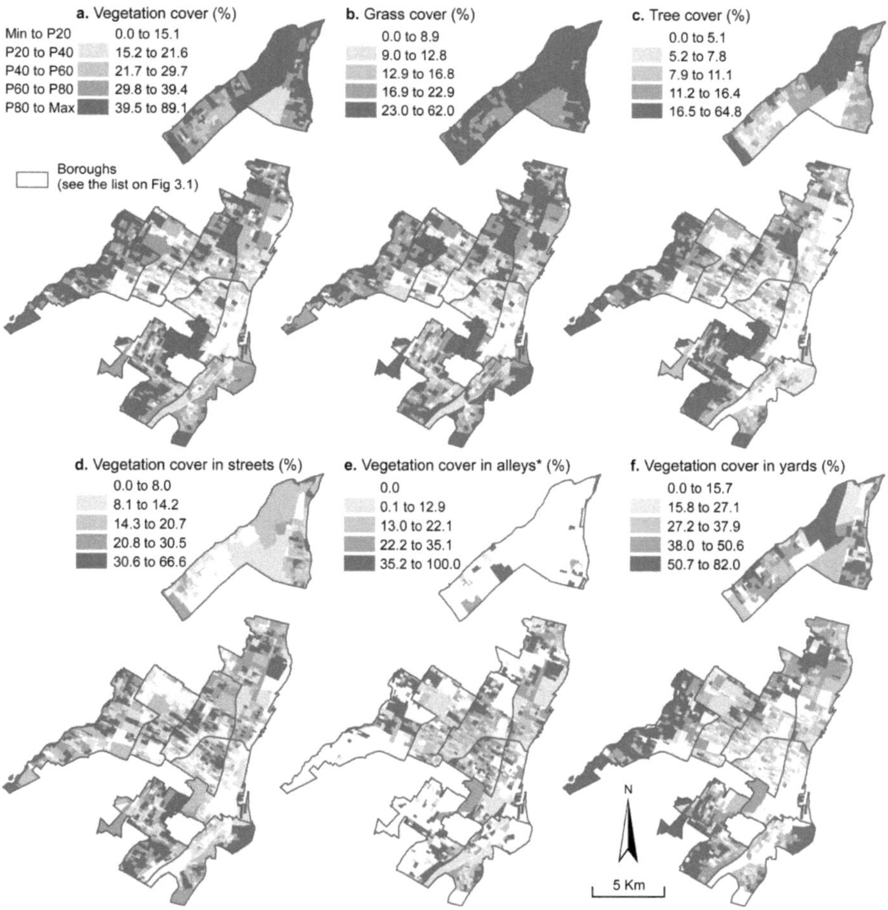

Figure 3.5 Vegetation, tree/shrub and grass cover (by dissemination area).

Ahuntsic-Cartierville, Rivière-des-Prairies—Pointe-aux-Trembles and Côte-des-Neiges—Notre-Dame-de-Grâce are the greenest boroughs as more than one third of their territory is covered by vegetation. In contrast, even with the inclusion of the regional park of Mont Royal, the less vegetated boroughs located in the urban core such as Le Plateau-Mont-Royal and Ville-Marie, have a vegetation rate lower than 17%. The low rate in Ville-Marie may be accounted by the fact that a part of vegetation is not detected by our method as it is masked by the shade of high-rise buildings here. We are also aware that both of the boroughs are densely built, indeed the most densely built in the city, which in turn leads to limited space available for vegetation.

However, the spatial variation of the indicators calculated for the entire DAs do not capture the spatial disparities of vegetation in streets, alleys and yards. First, the presence of vegetation is the highest in streets located in four boroughs of Côte-des-Neiges—Notre-Dame-de-Grâce, Rosemont—La Petite-Patrie, Ahuntsic-Cartierville, Mercier—Hochelaga-Maisonneuve (their rate being higher than 20% against 18% - the city's average rate). This indicator is the least in Ville-Marie (Figure 3.5d). Second, the vegetation cover in alleys is the highest in Plateau Mont-Royal, Rosemont—La Petite-Patrie and Mercier—Hochelaga-Maisonneuve (20% or more against 16.5%) (Figure 3.5e). Third, three boroughs make themselves stand out owing to the abundant vegetation in the yards (40% or more against 32%): Côte-des-Neiges—Notre-Dame-de-Grâce, Ahuntsic-Cartierville, Rivière-des-Prairies—Pointe-aux-Trembles (Figure 3.5f).

3.3.3 Relating Vegetation Indicators and Immigrant, Visible Minority and Low Income Populations

3.3.3.1 Correlation Analysis

The correlations between the vegetation indicators and the three sociodemographic variables reveal that in general, the immigrants and the visible minorities seem somewhat overrepresented in DAs exhibiting deficient vegetation (R^2 = -0.11 and -0.15) (Table 3.4). More precisely, they are more likely to be present in DAs exhibiting less vegetation in streets (R^2 = -0.27 and -0.29). Such correlations can be interpreted as an environmental inequity at a slight degree (the coefficients being relatively weak). However, the environmental equity issue seems more serious for the low income population with correlation coefficient values between -0.18 and -0.36.

These observations suggest that the environmental inequity related to vegetation exists in different forms according to the population groups: while for the immigrants and the visible minorities it is found in streets and alleys, for the incidence of low income it is clearly more elevated, be it in streets, alleys or yards.

Table 3.4. Correlations between the vegetation indicators in green spaces and sociodemographic data

		Immigrant	Visible minority	Low income
Total in DAs	Vegetation	-0.11	-0.15	-0.36
	Tree/shrub	0.00	-0.08	-0.28
	Grass	-0.18	-0.17	-0.35
Streets	Vegetation	-0.27	-0.29	-0.25
	Tree/shrub	-0.15	-0.21	-0.21
	Grass	-0.35	-0.33	-0.26
Alleys	Vegetation	**-0.08**	**-0.16**	**-0.22**
	Tree/shrub	**-0.03**	**-0.12**	**-0.19**
	Grass	**-0.11**	**-0.16**	**-0.21**
Yards	Vegetation	0.02	-0.03	-0.24
	Tree/shrub	0.10	0.02	-0.18
	Grass	-0.08	-0.07	-0.26

Bold values: significant at p<0.0001.

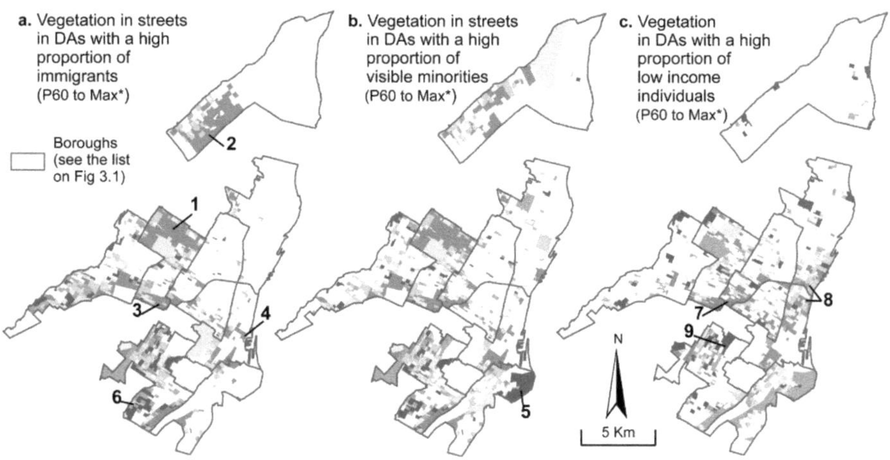

Number of DAs per class

Map	Problematic areas: Very high inequity	Problematic areas: High inequity	Non-problematic areas: No disadvantage/ no benefit	Non-problematic areas: Green benefits	Non-problematic areas: High green benefits	Total
a.	230 (32.5%)	156 (22.0%)	124 (17.5%)	104 (14.7%)	94 (13.3%)	709 (100%)
b.	219 (30.9%)	171 (24.1%)	134 (18.9%)	102 (14.4%)	83 (11.7%)	708 (100%)
c.	214 (30.4%)	170 (24.1%)	143 (20.3%)	111 (15.7%)	67 (9.5%)	705 (100%)

* (P60 to Max*): DAs where the percentage of a given population group is equal or higher than the percentile 60.

Figure 3.6 Problematic and non-problematic areas in terms of environmental equity.

3.3.3.2 Environmental Equity Mapping

Relying on the correlations mentioned above, we will focus on three maps depicting the relationships between: immigrants and vegetation in streets; visible minorities and vegetation in streets; and finally the incidence of low income and the total vegetation in the DA. On each map, the colored DAs represent areas where the proportion of a given group of population is high (P60 to Max) and the colour codes represent the levels of vegetation abundance. The first two maps, representing the immigrants and the visible minorities, show that the spatial distribution of the **very high inequity** areas is quite similar (Figure 3.6a and b). Overall 30% of the DAs displaying an elevated presence of these two groups are highly inequitable. They are located in Saint-Michel (**1**); Rivière-des-Prairies (**2**); Parc-Extension (**3**); and the downtown area (**4**). Moreover, the DAs with high green benefits are concentrated in Pointe Saint-Charles **(5),** Loyola **(6)** and scattered throughout Cartierville.

As for the low income population, three zones belonging to the **very high inequity** class are located along the railway from Parc-Extension to La Petite Patrie **(7)**, Centre-Sud and Hochelaga **(8)**, downtown and several DAs scattered in Le Plateau (Figure 3.6c). The number of **high green benefits** areas is much lower for low income population than when compared to the immigrant and visible minority populations (67 DAs versus 94 and 83). Only one zone, in the University of Montreal's campus, seems have a large number of **very high benefits** DAs **(9)**.

A few remarks concerning the urban fabric's characteristics are worth making in order to understand the presence of vegetation in the neighbourhoods just discussed. The **very high inequity** areas such as Saint-Michel were built rapidly after the WWII, a period during which cars were favoured as the main means of transportation, and hence have narrow sidewalks and lesser unpaved surface for vegetation. Old industrial zones and downtown are characterized by confined streets and scarce available space for gardens. There are also areas with **high green benefits** areas like Pointe-Saint-Charles which is a former industrial zone sectioned by railways and vacant lots where vegetation grows freely. Finally, in Loyola detached houses with gardens and well-vegetated streets contribute largely to the greenness of the neighbourhood.

3.4 Discussion and Conclusion

The use of satellite images together with an object-based classification allows us to evaluate and inventory different forms of vegetation – trees/shrub and grass – in the City of Montreal with an accuracy rate of nearly 75%. From this, we demonstrate that the proportion of vegetation varies significantly across the boroughs. The central districts have the advantage of richer vegetation in streets and alleys whereas the suburbs are greener thanks to the presence of vegetation in yards (for the most part grass). The results also prove that environmental inequity with respect to the three groups of population expresses differently depending on the type of green spaces, which is consistent to what is observed in other North American cities (Tooke et al. 2010; Landry and Chakraborty 2009; Iverson and Cook 2000).

Accounting for the mechanisms underlying the mentioned disparities warrants a multidimensional approach. First of all, the built environment influences the presence of vegetation as it defines the space available for the vegetation (Troy et al. 2007). Although we did not measure the built environment in a quantitative way, the very act of characterising the vegetation in streets, alleys and yards, allows us to evaluate qualitatively the impact that the built environment has on the amount of vegetation. Secondly, the neighbourhood legacy expressed by construction periods or tree protection and tree planting policies in the past can contribute greatly to the presence of current vegetation (Boone et al. 2010). A high percentage of renters and co-ownerships may also be an important factor that reduces the amount of vegetation (Landry and Chakraborty 2009). Which is why Heynen et al (2006) state that leaving urban greening programs to promoters and private interest groups could intensify environmental inequity. Finally, subjective factors are underlined by previous authors, namely lifestyle, social status and ethno-cultural preferences (Grove et al. 2006). Even though a study based on in-depth interviews also found that Mediterranean immigrants who arrived in Montreal before the 1970s tended to garden even on narrow lots (Routaboule et al. 1995), further research is required to verify if the rest of these arguments hold true for Montreal.

The results of this study may be of particular interest to decision-makers, urban planners and municipal services of parks and forestry. For example, local estimates of the amount of vegetation could be useful for allocating investment and promoting campaigns of greening in deficient areas. In the meantime, the areas with green benefits identified in this study could provide valuable insight for future urban planning policies seeking to improve access to green spaces for vulnerable populations.

Acknowledgments

This study had been funded by the Social Sciences and Humanities Research Council of Canada. We wish to thank Adrian Gould for his helpful proof reading.

References

Akbari H, Rose LS, Taha H (2003) Analyzing the land cover of an urban environment using high-resolution orthophotos. Landscape and Urban Planning 63:1-14

Bock M, Xofis P, Mitchley J, Rossner G, Wissen M (2005) Object-oriented methods for habitat mapping at multiple scales – Case studies from Northern Germany and Wye Downs, UK. Journal for Nature Conservation 13:75-89

Boone CG, Cadenasso ML, Grove JM, Schwarz K, Buckley GL (2010) Landscape, vegetation characteristics, and group identity in an urban and suburban watershed: why the 60s matter. Urban Ecosystems 13:255-271

Cackowski JM, Nasar JL (2003) The restorative effects of roadside vegetation: implications for automobile driver anger and frustration. Environment and Behavior 35 (6):736-751

Castonguay G, Jutras S (2008) Children's appreciation of outdoor places in a poor neighborhood. Journal of Environmental Psychology 29 (1):101-109

CIHI (2011) Urban physical environments and health inequalities. Canadian Institut of Health Information, Ottawa

Congalton RG, Green K (2009) Assessing the accuracy of remotely sensed data: Principles and pratices. Taylor & Francis Group, Boca Raton

CREMTL (2008) Mettre Montréal en alerte verte. Envîle Express, March 6th, p 3

Damm A, Hoster P, Schiefer S (2005) Investigating urban railway corridors with geometric high resolution satellite data. Paper presented at the ISPRS, Remote Sensing and Spatial Information Sciences, Tempe, 14-16 March

Delm AV, Gulinck H (2009) Classification and quantification of green in the expanding urban and semi-urban complex: Application of detailed field data and IKONOS-imagery. Ecological Indicators 11 (1):52-60

Freeman C, Buck O (2003) Development of an ecological mapping methodology for urban areas in New Zealand. Landscape and Urban Planning 63:161–173

Grove JM, Cadenasso ML, Burch WR, Pickett STA, Schwarz K, O'Neil-Dunne J, Wilson M, Troy A, Boone C (2006) Data and methods comparing social structure and vegetation structure of urban neighborhoods in Baltimore, Maryland. Society and Natural Resources 19:117-136

Heynen N, Perkins HA, Roy P (2006) The political ecology of uneven urban green space the impact of political economy on race and ethnicity in producing environmental inequality in Milwaukee. Urban Affairs Review 42 (1):3-25

Iovan C, Boldo D, Cord M (2007) Automatic extraction of urban vegetation structures from high resolution imagery and digital elevation model. Paper presented at the URBAN, GRSS/ISPRS Joint Workshop on Data Fusion and Remote Sensing over Urban Areas, Paris, 11-13 April

Iverson LR, Cook EA (2000) Urban forest cover of the Chicago region and its relation to household density and income. Urban Ecosystems 4:105-124

Jo HK, McPherson G (2001) Indirect carbon reduction by residential vegetation and planting strategies in Chicago, USA. Journal of Environmental Management 61:165-177

Kobler A, Dzeroski S, Keramitsoglou I (2006) Habitat mapping using machine learning-extended kernel-based reclassification of an Ikonos satellite image. Ecological Modelling 191:83-95

Kweon B-S, Sullivan W, Wiley A (1998) Green common spaces and the social integration of inner-city older adults. Environment and Behavior 30 (6):1998-1998

Landry SM, Chakraborty J (2009) Street trees and equity: evaluation the spatial distribution of an urban amenity. Environment and Planning a 41:2651-2670

Mallinis G, Koutsias N, Tsakiri-Strati M, Karteris M (2008) Object-based classification using

Quickbird imagery for delineating forest vegetation polygons in a Mediterranean test site. ISPRS Journal of Photogrammetry & Remote Sensing 63:237-250

Mathieu R, Aryal J, Chong AK (2007) Object-based classification of Ikonos imagery for mapping large-scale vegetation communities in urban areas. Sensors 7:2860-2880

Mathieu R, Pouget M, Cervelle B, Escadafal R (1998) Relationships between satellite-based radiometric indices simulated using laboratory reflectance data and typical soil color of an arid environment. Remote Sens Environ 66:17-28

McPherson G, Simpson JR, Peper PJ, Maco SE, Xiao Q (2005) Municipal forest benefits and costs in five U.S. cities. Journal of Forestry 103 (8):411-416

Myint SW, Gober P, Brazel A, Grossman-Clarke S, Weng Q (2011) Per-pixel vs. object-based classification of urban land cover extraction using high spatial resolution imagery. Remote Sensing of Environment In Press, Corrected Proof

Myint SW, Lam N (2005) A study of lacunarity-based texture analysis approaches to improve urban image classification. Computers, Environment and Urban Systems 29 (5):501-523

Qi J, Chehbouni A, huette A, Kerr K, Sorooshian S (1994) Modified soil adjusted vegetation index (MSAVI). Remote Sensing of Environment 48:119-126

Ridder KD, Adame V, Banuelos A, Brused M, Burgend M, Damsgaarde O, Dufek J, Hirsch J, Lefebre F, Pérez-Lacorzana JM, Thierry A, Weber C (2004) An integrated methodology to assess the benefits of urban green space. Science of the Total Environment 334-335:489-497

Routaboule D, Anselin V, Eveillard C (1995) "Le paysage de l'intérieur" ou expressions paysagères résidentielles dans l'île de Montréal. Société canadienne d'hypothèques et de logement, Montréal

Seeland K, Dübendorfer S, Hansmann R (2009) Making friends in Zurich's urban forests and parks: The role of public green space for social inclusion of youths from different cultures. Forest Policy and Economics 11:10-17

Thomas N, Hendrix C, Congalton RG (2003) A comparison of urban mapping methods using high-resolution digital imagery. Photogrammetric Engineering & Remote Sensing 69 (9):963–972

Tooke TR, Klinkenberg B, Coops NC (2010) A geographical approach to identifying vegetation-related environmental equity in Canadian cities. Environment and Planning B: Planning and Design 37:1040-1056

Toutin T, Chénier R, Carbonneau Y (2002) 3D models for high resolution images: examples with Quickbird, Ikonos and Eros. Paper presented at the ISPRS, Symposium on Geospatial Theory, Processing and Applications Ottawa, July 9-12

Troy A, Grove JM, O'Neil-Dunne JPM, Pickett STA, Cadenasso ML (2007) Predicting opportunities for greening and patterns of vegetation on private urban lands. Environ Manage 40:394-412

Walker G (2009) Exploring the multipe spatialities of enviromental justice. Antipode 41 (4):614-636

Witford V, Ennos AR, Handley JF (2001) «City form and natural process» - indicators for the ecological performance of urban areas and their application to Merseyside, UK. Landscape and Urban Planning 57:91-103

Yu Q, Gong P, Clinton N, Biging G, Kelly M, Schirokauer D (2006) Object-based detailed vegetation classification with airborne high spatial resolution remote sensing imagery. Photogrammetric Engineering & Remote Sensing 72 (7):799-811

Zhang X, Feng X, Jiang H (2010) Object-oriented method for urban vegetation mapping using IKONOS imagery. International Journal of Remote Sensing 31 (1):177-196

Zhang Y (2002) Problems in the fusion of commercial high-resolution satellites as well as Landsat 7 images and initial solutions. Paper presented at the ISPRS, GeoSpatial Theory, Processing and Applications, Ottawa, 9-12 July

Zhou W, Troy A (2008) An object-oriented approach for analysing and characterizing urban landscape at the parcel level. International Journal of Remote Sensing 11 (10):3119-3135

Chapter 4

Geodemographics and Environmental Justice in Montreal: Exploring Socioeconomic Position Through Geodemographic Segmentation

Tom Weatherburn[1] and Daniel Naud[2]

[1] Master of Spatial Analysis at Ryerson University
[2] Doctorant, Département de Géographie, Université de Montréal

Abstract

Developing a deep and thorough understanding of environmental justice in urban areas depends on the clarity through which researchers and practitioners can assess the socio-demographic structure of a city. Geodemographic segmentation systems provide a classification technique through which a multitude of social and economic indicators are used to profile and map neighborhood characteristics. Although geodemographic segmentation has been applied to a diverse body of commercial and academic research, the development of a specific application in the context of environmental justice assessment may provide valuable insights for the case of Montreal. A dataset was compiled to represent Socioeconomic Position (SEP) using age, sex, ethnicity, language, income, occupation, education and urbanity indicators. The data set was segmented using a hierarchichal ascending classification (HAC) cluster analysis to show statistically similar regions of Montreal at the scale of the Dissemination Area. The mapping and visual interpretation of the resulting geographic clusters showed distinct patterns of socio-demographic distribution across the aforementioned themes. Although no regression analysis has been completed, the cartographic representation of the socio-demographic structure of Montreal will provide guidance for a more developed statistical analysis.

4.1 Introduction

The primary concern of the environmental justice movement is the examination and rectification of disproportionate levels of exposure to environmentally hazardous, health-degrading pollutants by low-income or ethnic minority populations (Apelberg et al. 2005). Developing an understanding of environmental justice in urban areas depends on the clarity through which researchers and planners can interpret the socio-demographic structure of the urban population. As such, the segmentation of Montreal was completed with the distinct objective of visualizing the socioeconomic variation in the population, based on location, to address basic questions relevant to an exploratory assessment of environmental justice. Are ethnic minority populations clustered in specific spatial patterns in Montreal? Is wealth concentrated in a similar way? Are there other patterns that characterize the socio-spatial organization of Montreal residents? Geodemographic segmentation systems provide a classification technique through which a multitude of social and economic indicators are used to profile and map neighborhood characteristics. By using an established, yet general, framework of indicators for assessing socioeconomic position (SEP) (Galboredes et al. 2006) and the resultant representations of socioeconomic distribution certain inferences and preliminary questions can be answered regarding populations that may be deemed "at risk" in the context of environmental justice. Although the framework for such assessments is complex, the analytical framework is built on a foundational understanding of socioeconomic geography in urban regions such as Montreal.

4.2 Environmental Justice, Socioeconomic Position and Geodemographics

The notion of environmental justice was defined in 2005 by Apelberg et al. as "the movement concerned with inequities in the distribution of adverse environmental and health consequences of industrial activities and environmental policies." While they note that the basis of this movement began with Brown (1995), the disproportionate "burden" of exposure to environmental pollutants on communities with low-income or a high percentage of visible minorities was studied as soon as the second half of the 1980 decade, as observed by Maantay (2002: 161). Therefore, the Environmental Protection Agency in the United States has developed a working mission by stating that "no group of people, including a racial, ethnic, or a socioeconomic group" should be disproportionately affected by "industrial, municipal, and commercial operations or the execution of federal, state, local, and tribal programs and policies."

Applying broader definitions of environmental justice becomes increasingly complex as the task of interpreting status or "position" within social structures emerges. Taylor (2000) discusses the rise of the environmental justice paradigm as a useful tool to examine the very pertinent idea of socioeconomic position and its implications on class, power and political participation. Taylor argues that, where race and income levels are important descriptive characteristics of unjustly treated populations; a clear definition of SEP should aim for the incorporation of gender and a more broad interpretation of class. Brulle and Pellow (2006) also argue that because environmental inequality is a product of socio-political hierarchy, the examination of race, gender and class cannot be isolated when trying to understand this structure. Therefore, economic, racial, and social indicators should be used collectively to assess a group's relationship to power and decision-making. Ultimately the concept of SEP is defined through a multitude of socioeconomic indicators. These indicators reflect a group's relationship to power in society by illuminating barriers to the effective advocacy of collective interest. Rast (2006) provides a compelling example by examining the participation of minority groups in planning initiatives such as urban revitalization and smart growth. Rast argues that although planning and policy development often focus on minority groups and low-income communities, white, suburban, middle-class populations overwhelmingly bias the dialogue. Rast concludes that environmental justice may be an effective platform for the greater participation and engagement of inner-city minority communities (Rast 2006).

In Canada, research has been somewhat limited although examples of environmental justice research have taken place. Buzzelli et al. (2001) incorporated a broad definition of environmental justice into a site-specific analysis. They examined the spatial distribution of dwelling values and unemployment rates in Hamilton Ontario, noting that a clear pattern of differentiation existed in relation to toxic air pollutants. They note, however, that categorizing such observations becomes difficult as the decline in correlation between social markers and pollution levels is largely attributable to postfordist economic reorganization as opposed to specific policy initiatives, designed to address inequalities. More recently, the analysis of satellite imagery and census data in some Montréal's neighborhoods showed a weak, but significant negative relationship between green spaces and low-income households (Pham et al. 2010). A comparison with American and developing countries cities revealed that this inequality is much lower in Montréal. In 2005, Sénécal, Hamel and Vachon explored the relationship and spatial structure of several urbanity and socioeconomic variables. They found out three dominant structures, based on 1) built density, distance from the city center and vegetal cover; 2) income and home ownership; and 3) proximity to jobs. On the contrary of Brulle and Pellow (2006), they did not add ethnicity related variables. Another Montréal study (Smargiassi 2008) revealed that financially disadvantaged households are not necessarily more exposed to heat and air pollution, but they are more at risk, considering they sustain negative effects from pollution as well as poverty.

In the context of Environmental Justice, theories on defining SEP are closely aligned with much of the research being done in epidemiology. As such Galobardes et al. (2006) have been used to explain the utility of some of the SEP indicators utilized.

Education can be thought of as Knowledge based assets and is thought to be an important indicator of socio-economic position. Measuring education can be done through years in school or as the highest level of education achieved. According to Galobardes et al. (2006), Education captures the transition from childhood (received) SEP to adulthood (own) SEP and largely determines future employment and income. Education is reflective of the material and intellectual resources made available at the family origin, usually ends in young adulthood. Higher levels of education result in a more advanced cognitive functioning, and can make people more receptive to new health or environment related information. Education can have a direct influence on accessing strategic or political resources that enable participation in planning and other actions that influence their reality.

Housing in general is often thought of through housing tenure, housing condition and housing amenities. Measurement of housing status is most commonly done through housing tenure -ownership of property versus renting or public provision. Based on region tenure statistics can be complimented by housing type to indicate density and land value as well. Housing is largely a material indicator, acting as a key component of wealth and outgoing income. Although housing is fundamental to assessing SEP and environmental justice, it can be complex and geographically varied based on environmental and economic determinants.

Income is the most commonly used indicator of socioeconomic position and is the most direct measure of material resources. Income levels are also cumulative over life course and can be the most volatile over short periods of time. The complexity of income is that it can influence SEP in different ways. To be specific, money itself is not a direct influence on health (Galboredes et al. 2006: 10) or environmental justice, thus interpreting the conversion of money into "health enhancing commodities and services via expenditure," or in this case, expenditure enabling relocation of one's self or of environmental hazards, that can improve environmental influence on health is more important. It is the way money is used to increase social standing and esteem, thus encouraging engagement and social participation that can have direct impacts on environmental justice issues.

Occupation based indicators of SEP are widely used. Occupation becomes a useful indicator when looking at Max Weber's definition of SEP as a 'market position' in society (Galboredes et al. 2006: 8). Occupation often relays a person's relationship to employers and employees as a common proxy for a more generalized social stratification. In some cases, it is useful when viewing a person's position as relative to means of production and exploitation. As with many of these indicators, as a sole indicator, the influence or implications of occupation are difficult to isolate from other. However, generally, occupation contributes to understanding access to material resources or access to decision-making in planning or health initiatives.

Ethnicity, as discussed by Brulle and Pellow (2006), can't be left behind of such a study, considering the literature often associates visible minorities with highly disadvantaged areas, even in Montréal (Leloup and Apparicio 2010: 190-1). On the same idea, the geolinguistic division of Montréal is to be taken into consideration, since it is historically a structure with a high degree of influence over socioeconomic issues.

Overall, in the context of this study we propose to analyze the spatial structure of these different criteria, using geodemographic segmentation. The latter is an effective tool for the research and planning of public services and urban policy development. Geodemographic segmentation is a tool to classify predefined spatial units based on socioeconomic data in an attempt to visualize demographic distributions in the context, for example, of SEP in urban settings. Longley (2005) outlines the utility of geographic information system (GIS) based geodemographics for public service delivery by stating "GIS makes it possible to devise practical, rational measure of local service demand as well as facilitating transparent assessment of current performance levels relative to locationally sensitive levels of need." The focus of much commercial geodemographic segmentation on profiling 'lifestyle' is analogous the idea of using SEP as a useful representation of social hierarchy. The combination of socioeconomic data with geographically referenced data allows SEP to be defined not only in the abstract but also spatially. The potential then exists for planning and policy development to be informed in a very meaningful way by the marriage of these data structures.

4.3 Data and Methods

Methodologically, the process of constructing a geodemographic segmentation system consists of several steps. The data was collected from the Statistics Canada census database, at the Dissemination Area (DA) level, and prepared through the amalgamation of certain variables and the conversion of values to percent of total population. Amalgamating variables was conducted by summing the values of several variables into logical groups so as to simplify the analytical process. For example, management, business, finance and administration, natural and applied sciences, health, social science, education, government service and religion, were amalgamated into the white-collar group. Table 4.1 lists all analyzed variables, as well as their descriptive statistics. Missing data usually refers to uninhabited areas (i.e.: parks). The descriptive statistics reveal normal distributions in most cases. *Visible Minorities* and *Immigrant Population* variables have relatively high, positive skewness values, which are explained by a large number of spatial units having low proportion of immigrant population, while a few spatial units have a high proportion of immigrant population.

Table 4.1. Socioeconomic Indicators, Montréal, 2006 (N = 3115).

Variables	Missing (N)	Mean (%)	Median (%)	Std. Dev.	Skewness	Kurtosis
Owned dwelling	28	41.9	34.4	29.3	0.7	-0.6
Rented dwelling	28	57.8	65.4	29.3	-0.7	-0.7
French first lang. spoken	16	61.6	66.3	26.2	-0.4	-1.2
Unemployed	16	8.8	7.7	6.3	1.2	2.5
White collars	16	56.6	57.1	15.3	-0.2	-0.2
Blue collars	16	15.4	14.3	9.9	0.6	0.0
Service workers	16	23.4	23.1	8.3	0.4	0.9
Drive to work	16	51.5	51.8	17.6	0.0	-0.6
Public transport to work	16	32.7	32.3	14.4	0.1	-0.4
Walk to work	16	7.7	5.6	8.6	2.4	8.5
No High School Degree	16	14.0	12.1	10.8	0.9	0.8
High School Degree	16	19.3	18.6	8.6	0.5	0.4
College Degree	16	16.0	15.4	6.9	0.5	0.7
Bachelor Degree	16	18.8	17.9	9.9	0.4	-0.1
Graduate Studies Degree	16	3.0	2.7	3.5	1.4	2.9
Visible Minorities	16	21.7	17.2	23.3	13.9	420.8
Immigrant Population	16	27.2	24.8	23.8	17.2	527.5
Low Income (<$30K)	38	22.3	20.0	16.0	0.8	0.4
Lower Mid Income ($30K-$70K)	38	41.5	42.1	15.6	-0.1	-0.1
Upper Mid Income ($70K-$100K)	38	16.5	15.6	10.7	0.5	0.0
High Income (>$100K)	38	19.7	14.3	19.0	1.3	1.4

Considering the large size of the database, in terms of observations (3115 ADs) as well as variables (21 variables), two data reduction techniques were used in this exploration of Montreal. A Principal Component Analysis (PCA) was run to reduce the number of variables to interpret, given their large number and the collinearity between several of them (i.e.: variables *Owned Dwelling* and *Rented Dwelling* are negatively correlated). In a PCA, correlated variables are transformed into new variables called *Components* or *Axes*, which are independent or uncorrelated with the other new components (Rogerson 2006: 257). Factor analysis such as this one has already been applied in environmental justice studies in the Montreal metropolitan region, at the census tract level, to reveal unforeseen or socio-spatial structures (Sénécal, Hamel and Vachon 2005).

The segmentation process, involved two different cluster analysis methods. The objective of cluster analysis is to group together similar spatial units, thus limiting the number of observations and facilitating the interpretation. The process can be hierarchical or nonhierarchical (Rogerson 2006: 265). The former starts with n clusters, n standing for the total number of observations. For each step, the process merges one cluster with another, so it remains n-1 clusters. The process stops when no more merges can be done: the last cluster contains all n spatial units. It is called hierarchical, because no merger can be undone in subsequent steps. The process can also be non-hierarchical, in which the analyst has an idea of the number of relevant classes. The process starts with a known set of seed points (centroids) and for each step, an observation is merged in the cluster with the centroid with which it is the most similar, centroid which is then recalculated to reflect the new addition. The process continues until no observation can be moved without increasing its dissimilarity with a seed point. In both methods, the resulting clusters are to be defined based on variable interpretation and ancillary knowledge, in order to become the resulting socioeconomic groups.

While the latter process is the fastest, it also required the analyst to have an *a priori* idea of the total number of clusters to achieve. Since this paper is more about the exploration of geodemographic segmentation and there's no finite and explicit number of clusters to analyze, we will instead resort to a hierarchical method, which has the advantage of offering tools to find the optimum number of classes. This segmentation is accomplished with a Hierarchical Ascending Classification (HAC). The latter has already been applied to socioeconomic spatial segmentation to explore the Montreal's region (Leloup and Apparicio 2010; Apparicio and Seguin 2006). Squared Euclidean distance has been used to measure the distance between two clusters and Ward criterion as the aggregation method (minimization of the within-group variation; the Ward's criterion picks up the observation that adds the lesser within-group variance to the cluster).

Finally, only on an exploratory basis, we compared proportions of several health issues in the population, according to the previous segmentation. Health issues will serve as a proxy to environmental issues, as a consequence of soil toxicity and bad air quality. A simple analysis of variance (ANOVA) will compare the variation of the data between clusters and within clusters. A significantly larger variation between groups will indicate that their differences are not caused by random fluctuations. This database has been compiled during the winter 2010 by students of the Diploma in Environmental Assessment (Department of Geography, Planning & Environment) at Concordia University (Montreal), from 2001 and 2006 censuses data and the Atlas Santé (Carrefour Montréal). Since the dataset is at the health sectors (CLSC) level, the data was spatially transposed at the DA level, using ArcGIS and the spatial join feature, which has obviously a serious limitation.

4.4 Limitations

It is very important to understand that the system of geodemographic segmentation is a largely generalized representation of socioeconomic characteristics. Throughout each of the cluster groups a degree of heterogeneity exists that cannot be ignored. Although in many cases a dominant characteristic emerges, there are inevitably cases that are far less clear. However, the assumption is made that the given spatial unit (DA) is characterized by certain attributes more than others. The degree to which this occurs is unique and a potential avenue for further analysis and visualization. Also, the data from the Health sectors (CLSC) is aggregated at larger areas scale than the DA level, so its application at this scale contains certain assumptions. Although interpolation techniques could have been used to obtain a more accurate depiction of the distribution of health issues, the data was left as is. Based on the exploratory nature of the analysis, it will nonetheless provide the analysts with a good indication of the utility of the segmentation process as a tool for environmental justice evaluation.

Continually, the segmentation outputs are very much proportionate to variable selection processes. Variable weighting has not been used in this methodology, but it is possible that some indicators are more important in understanding socioeconomic distributions in the context of environmental justice and health equity, for examples income and ethnicity. There are however, alternative additions that could illuminate an aspect of socioeconomic spatial structure in Montréal that has not been captured by the current dataset, as age group and number of children.

4.5 Results

In the following results, we will outline the execution of the PCA, and show the reduction of the total number of original variables to only a few principal components (or axes). Based on the loading of each component, we will then show how we merged groups of similar DAs, with the objective to create a simple and easy to interpret geodemographic segmentation of the island of Montreal, using a HAC.

In Table 4.2, the *Total* column is the *eigenvalue*, which is the sum of the squared correlations between the factor and the original variables. The first component has an *eigenvalue* of 6.66 and is explaining 31% of the variance. Only the four firsts components were selected, the others had lesser contribution to the explanation of variability in the dataset. This means that most of the variability in the original dataset can be explained by only four synthesized variables (even though the fourth one, as we will show, is harder to interpret).

Table 4.3 provides a mean to interpret the four components, by stating the correlations between a component and all variables. The variables scoring higher within the components were grayed to help the reader identify them. The higher the *loading*, the more correlated with the component the variable is. In other words, those grayed variables were combined to form the actual component. A negative loading tells a negative correlation between the original variable and the component. *Component 1* is associated with housing, commuting and income. *Component 2* is associated with occupation, education and, though less than *1*, income. *Component 3* is associated with French language, immigration and visible minorities. Finally, *component 4* has lower loadings and shows some redundancies with components 1 and 2, but is associated with education and income. Each DA is given a score on all components, defining its relation with it. For example, a DA scoring negatively on component 1 will mostly be composed of owned dwellings, where the inhabitants, more probably wealthy than poor, drive to commute to work. The following figure shows the resulting maps of the DA scoring on all four components, as well as their synthesized name.

Table 4.2. Variance explained by each component.

	Extraction sums of squared loadings			Rotation sums of squared loadings		
Component	Total	% of variance	Cumulative %	Total	% of variance	Cumulative %
1	6.7	31.7	31.7	4.6	22.0	22.0
2	3.1	14.6	46.3	4.6	21.9	43.9
3	2.3	10.0	57.3	2.5	11.7	55.6
4	1.2	5.6	62.9	1.6	7.4	62.9

Extraction method: Principal Component Analysis.

Table 4.3. Factor Loadings.

Variables	Components			
	1	2	3	4
Owned dwelling	-0.19	0.01	0.06	-0.03
Rented dwelling	0.19	-0.01	-0.06	0.03
French first lang. spoken	0.07	0.08	-0.36	0.07
Unemployed	0.03	0.01	0.11	0.16
White collars	0.04	-0.18	-0.06	-0.06
Blue collars	-0.08	0.20	-0.01	0.09
Service workers	0.02	0.09	0.08	-0.07
Drive	-0.25	0.10	0.00	0.11
Public transit	0.22	-0.05	0.06	-0.21
Walk	0.13	-0.12	-0.07	0.13
No High School	-0.05	0.17	-0.07	0.22
High School	-0.09	0.17	-0.01	0.00
College	0.11	0.02	0.09	-0.63
Bachelor	0.06	-0.21	0.05	0.02
Graduate Studies	0.00	-0.13	0.09	0.09
Visible Minorities	0.02	0.00	0.33	-0.04
Immigrant Population	0.00	-0.04	0.39	-0.06
Low Income	0.10	-0.01	0.05	0.19
Lower Mid Income	0.08	0.12	-0.04	-0.18
Upper Mid Income	-0.01	0.03	-0.01	-0.32
High Income	-0.14	-0.10	-0.01	0.17

Extraction Method: Principal Component Analysis.
Rotation Method: Varimax with Kaiser Normalization.

To reflect the scorings on a component which can be positive or negative, a diverging symbology has been used. The darker tones show negative relationship with the component, the lighter tones show a positive one, and the neutral color (almost white) shows a weak relationship. A short description of each end of the axis accompanies the legend.

The first component, called Dwelling and Income, shows that higher income households, living in single family houses, are located in the western and eastern

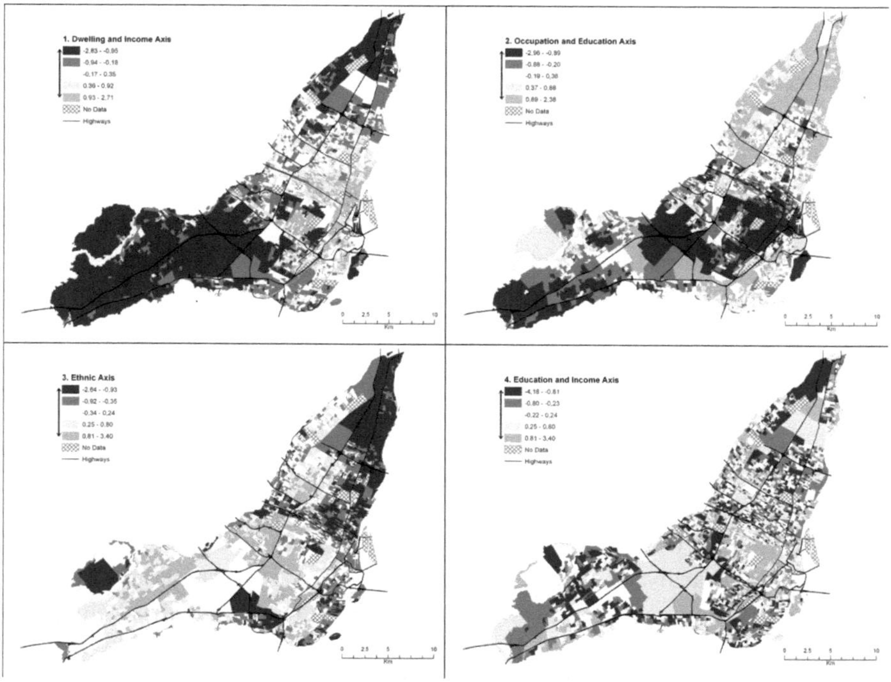

Figure 4.1 Socioeconomic axes from the PCA, Montréal, 2006.

parts of the island, as well as around the Mont-Royal Park. We can note though that they are clearly more prominent in the western part. On the other hand, households with rented dwelling and lower income are located in the central part of the island, more specifically in the southern area than the northern one. The pattern is clear and understandable, since the availability of single-family houses is larger in the suburbs than in the inner-city.

The second component is called Occupation and Education, since it was highly correlated with variables associated to these themes. White collar populations, with a strong education profile, are more closely related to the western area of Montreal, especially in the inner-city, around the Mont-Royal Park and in the West Island. The blue collars populations are to be found in higher proportions in the eastern part of the island. Also, note that the South-West shows a weaker relationship with the component (low scores), which reveals diversity on this axis.

The third component is related to ethnicity, scoring high with the language, the immigration and the visible minorities' variables. This map shows a clear structure, with French speaking populations with low ethnic diversity are located in the south-eastern part of the island. Some pockets are also found following the orange metro line on the north-south axis in the central part of the island, in Verdun and in Lachine. The more ethnically diversified areas are located in the north-eastern part of the island and in most of its western area. Note that several AD of the West Island score very low on the axis, thus revealing a sociodemographic structure that is harder to locate on such an axis. The West Island has lower proportion of French speaking populations, but also has a lower ethnic diversity, so several ADs could not get a high score on the axis, neither positively or negatively.

Finally, the fourth component doesn't show any clear spatial structure. In fact, even the interpretation of the loadings (Table 4.3) was troublesome, because a positive score on such axis could mean a higher proportion of the population without a high school degree (0.22), but also low income (0.19) or high income (0.17). It is left for the sake of transparency, but unfortunately, no relevant discussion on this axis could be made.

Some structures are emerging from these maps and the merging of these DAs could without doubt lead to suitable geodemographic segmentation. As discussed, a Hierarchical Ascending Classification (HAC) was run on the four previous components. The HAC output gives some help to determine the optimum number of clusters to be analyzed and mapped. Since the size of the dataset is very large, the dendrogram figure and the agglomeration schedule for hierarchical clustering table are left out of the paper. But both showed that the largest jump in the distance coefficient lead to a four clusters segmentation.

Table 4.4. Four-cluster solution (mean scores).

Component	Cluster			
	1	2	3	4
1	0.73	0.44	-0.47	-1.35
2	-0.57	0.58	0.93	-0.62
3	-0.45	1.00	-0.71	0.37
4	-0.21	0.34	0.12	-0.15
DA Count	1126	729	642	618

In the previous table, we can compare the mean scores of each cluster on the four principal components, so it is possible to define the four clusters, by referring to the loadings of each variable in Table 4.3. In the following definitions, you will find between brackets the component number the variable is associated with. The *first cluster* has a high positive mean score on the first component and lower negative scores on the other three. It means the first cluster has high proportion of households in rented dwellings[1] (1), composed of white collars[2] (2) with a high education profile (2 and 4), mostly French speaking (3), with a large range of income (1, 2 and 4; though component 1 has the largest contribution to the segmentation process, so it should tend to lower income rather than very high). The *second cluster* scores positive on all components, but mostly on the second and third. This means it is composed mostly by households in rented dwellings (1) with lower income (1, 2 and 4), higher proportions of blue collars and service workers (2) with a high school degree (2) or no degree at all (4), high proportions on non-French speakers (3), visible minorities (3) and immigrants (3). The *third cluster*, with positive score on components 1 and 3 and negative score on 2 and 4, reveals French-speaking (3) suburbanites' households (1) in low ethnic diversity areas (3), driving to commute (1), in the upper income range (1), but still mostly blue collars and service workers (2) with high school degree (2). Finally, the fourth cluster has a very high negative score on component 1, negative scores on 2 and 4, and a positive one on 3. This means a high proportion of non-French (3) owners (1) with white collar occupations (2) and high income (1, 2 and 4), highly educated (2 and 4), in areas with some ethnic diversity (3). This being said, the most interesting aspect of exploring sociodemographics through geodemographic segmentation is the mapping of these four clusters, shown in Figure 4.2.

As it have been said previously, it is very important that analysts as well as the readership keep in mind that this segmentation is based on generalization: only four clusters can't reveal the complexity of such a large region. As it is always the case with spatial analysis, a least a general geographical knowledge of the area is required for drawing the most benefit from the results. Having this in mind, the cluster analysis is powerful enough to give a lot of information about a region. We clearly see a structure emerging from this segmentation, with two very distinct clusters on both tips of the island (clusters 3 and 4), while two clusters share the central area (clusters 1 and 2). The higher socioeconomic profiles can be found in the West Island, but also surrounding the Mont-Royal Park, in Outremont, Westmount and Hampstead. We know from the geolinguistic division of Montréal that the West Island is mostly English-speaking, which explains why this cluster scored positively on component 3. The other end of the island is inhabited mostly by French speaking middle-class suburbanites. We also see some DAs from this cluster in the South-

[1] Since the DAs associated with Cluster 1 usually score high on Component 1 (mean = 0.73), we refer to Table 4.3 to find a negative relationship with owned dwellings (-0.19) and a positive one with rented dwellings (0.19).

[2] On the contrary, in this case DAs associated with Cluster 1 score low on Component 2 (mean = -0.57). By referring to Table 4.3, it is understood that a positive score means a negative relationship with White collars (-0.18) and positive one with Blue collars (0.20). Since Cluster 1 has a negative score, the relationships are reversed, so there is a positive relationship with White collars and a negative relationship with Blue collars (as well as a positive relationship with Bachelor and Graduate Studies, -0.21 and -0.13, respectively).

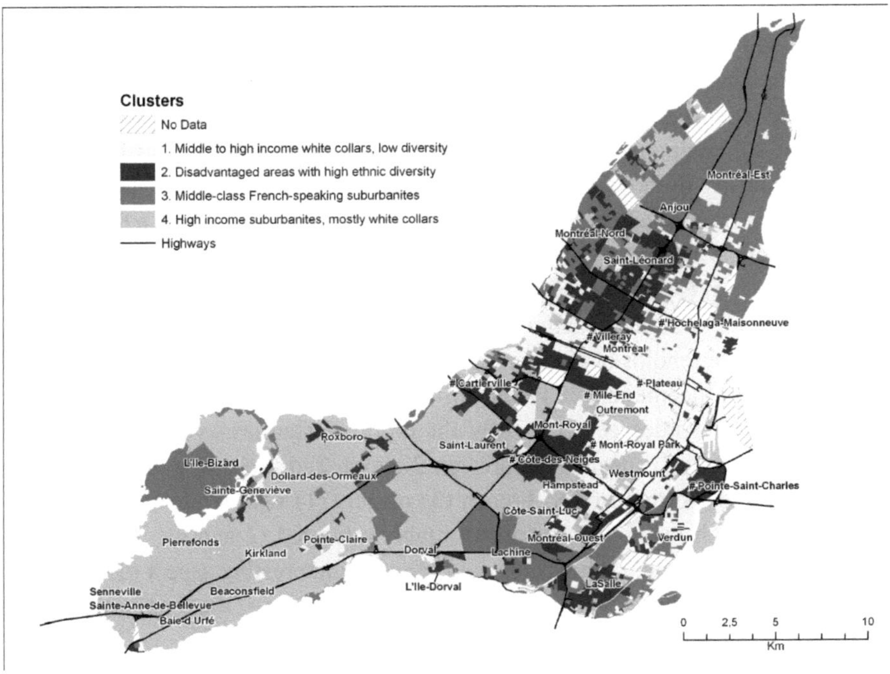

Figure 4.2 Socioeconomic segmentation of Montréal, 2006, with labelled municipalities and relevant neighbourhoods (with #).

West (Lachine, Lasalle and Verdun). We recognize some industrial land use within these areas, which might explain the blue collar occupations. As for the cluster called *Disadvantaged areas with high ethnic diversity*, it is important to recall that even though there's tendency towards low income and education, all DAs within it might not necessarily be socioeconomically disadvantaged. We find it around Côtes-des-Neiges, Cartierville (east of Saint-Laurent), Saint-Leonard, Montréal-Nord, Pointe-Saint-Charles and Lasalle (in the South-West). This cluster is more scattered than the others, which could reveal (though it is not the case with this exploration) some socioeconomic differentiation between the several areas, in terms of ethnic origin or income, for examples. And the last cluster, composed of middle-class white collars in rented dwellings, is to be found around the Mont-Royal Park and East of the Saint-Laurent Boulevard. While it is not displayed on the map, it is very interesting to note how well this cluster fits to the subway network. This cluster covers districts such as Villeray, Plateau-Mont-Royal, Mile-End, Centre-Sud, parts of Westmount and Outremont, and progressively stops in Maisonneuve.

Finally, to align the segmentation with environmental justice issues, we compared these four clusters to several health statistics. In the following table, we can compare the actual mean values of the health statistics of all four clusters[3].

[3] An analysis of variance (ANOVA) was run to compare the means of the four clusters on all health

Table 4.5. Clusters' mean values by health issues.

	1	2	3	4
Cardiac	0.12	0.14	0.14	0.12
COPD (Pulmo)	0.47	0.42	0.51	0.32
Underweight	0.35	0.37	0.37	0.36
Premature Birth	0.44	0.45	0.46	0.45
Retardation	0.54	0.55	0.51	0.47
Mortality	0.10	0.09	0.10	0.08
Respiratory Mort.	0.01	0.01	0.01	0.01
Cancer	0.07	0.06	0.07	0.06
Lung cancer	0.03	0.02	0.03	0.02

It is important to remember that clusters 1 and 4 were more financially advantaged than 2 and 3; that cluster 2 is the most ethnically diversified and cluster 3 is located in known industrial zones (i.e.: Lachine, Montréal-Est, Montréal-Nord). In dark gray are the largest values and in light gray, the lowest, on each variable. Without any deep analysis, we note that cluster 4 is by far the less affected by diseases and health issues. It is most often the cluster that scores the lowest proportion of its population affected by diseases. Interestingly enough, cluster 2 could have been thought to be in the most disadvantageous situation, considering its financial and occupational context. However, it should not be thought of as an immune population, considering it still scores high on several variables (cardiovascular problems, underweight, and retardation) that can have environmental causes. Cluster 3, mostly composed of French blue collar and service workers, is by far the most at risk population on the island of Montréal. The proximity of cluster 3 to several industrial zones surely increases its susceptibility to health issues related to environmental causes. Cluster 1, mostly composed by French white collars living close inner-city, is also an interesting case. It scores low on birth health issues, but high on mortality and cancer.

variables and determine if they were significantly different. If means are not different between clusters, they do not answer appropriately the assumption of geodemographic segmentation: no segmentation would give the same results. Fortunately, they were all significantly different, with a threshold of 1%. Thus, it gives us confidence in the geodemographic segmentation method used.

4.6 Discussion

As an exploratory endeavor, the geodemographic segmentation of socioeconomic variables in Montreal has provided an intriguing point of departure for a more rigorous statistical analysis of environmental justice and health equity. A quick analysis of this geodemographic segmentation reveals a distinct socioeconomic spatial structure, which could prove to be relevant in assessing environmental equity. Effectively, the most disadvantaged populations are in the northeastern part of the island, but are also scattered in the central part, where we know they are in close proximity to industrial land use. By mapping the distribution of socioeconomic components and creating a classification system to agglomerate groups of population based on common traits, it has been shown that where a person lives will describe a great deal about who they are. Dissecting the distribution of one variable, such as income, by other variables, such as ethnicity, education level or housing type showed that people are not only grouped or divided along economic lines, but also, to different degrees, along cultural and structural lines.

It can be said that one of the most important factors shaping the spatial distribution of clusters is the level of ethnic diversity and the presence of visible minority populations in a given region. This spatial concentration of ethnically organized clusters is a commonality in many North American cities (Balakrishnan and Grimyah 2003; Pamuk 2004) and Montreal is no exception (Apparicio, Leloup and Rivet 2006). It is important to note that a clustered ethnic spatial structure does not refer to such dense concentration as to be seen as ghettoization. Therefore, such concentration should not be understood *a priori* as a negative phenomenon. As Hou and Picot write (2004: 14), residential segregation contributes to the stability of ethnic identity, but also education, religious and community institutions required for active social interaction. Although French and English speaking populations are the majority in most areas, it became clear that some areas contain pockets of concentrated ethnic communities that are scattered across the central region of the island. While the fourth cluster scored high and positively on the third component, we know from the mapping that this was mostly due to the English speaking characteristic of the population. The third cluster was the best one to reveal this pattern of concentration and scattering.

Beyond ethnic clusters outlined above, the most predominant distinctions between cluster groups, the remaining of which are largely of European descent, are based on income (and the associated education, occupation and housing type), as was expected in Montréal. We do find two clusters with high socioeconomic profiles (3 and 4), differing largely in terms of language and housing status (this might be a cultural trait, that Montréal French-speaking populations tend to live in rented dwellings). Two clusters have lower socioeconomic profiles, mostly located in the eastern part of the island. We do find them in known residual areas, where industries are or were dominant. This seems coherent with the discussion made by Sénécal, Hamel and Vachon (2005: 38), which proposes that the Montreal socio-economic spatial structure is still largely a tributary of urban forms dating back to its industrial era.

It is interesting to note that economic factors might not be the strongest to determinant of susceptibility to environmentally caused health issues. While it seems to have an influence over the fourth cluster in the Western part of Montréal, it seems to be less relevant with the three other clusters. The first cluster, even though is comprised of households with sufficient incomes, is affected by serious health issues. On the other hand, the second cluster, located in more socioeconomically disadvantaged areas and with higher proportion of visible minorities' populations, is less affected by the studied diseases. While this does not contradict totally the observation made by Brown (1995), it brings to the environmental justice field a new input on the importance of geodemographic segmentation. Effectively, analysts should be more interested in the socioeconomic spatial structure which can in fact destabilize predetermined or conventional thoughts on the "burden" of low-income populations and visible minorities in segregated areas. The present study revealed that in Montréal, this burden might in fact be more importantly caused by occupational and income factors, that are rooted in education levels, rather than ethnicity ones.

In regards to the ultimate question of environmental justice, there are several dimensions that may be relevant in the case of Montreal. Overall it has been shown that ethnic minorities are spatially clustered throughout the central and eastern regions of the island, while a major division between French and English populations occurs from east (French) to west (English). The most affluent clusters are largely in the central to western portions of Montréal. Therefore it is reasonable to say, the context of the mean values for environmental health afflictions, that the populations at most risk for environmental injustice reside, generally, in the north central and eastern areas of Montreal, but also in some industrial pockets in the south-west area of the island.

4.7 Conclusion

The process of policy development often requires the use of simplified categorization, even if one has to admit some generalization, to allow quick and opportune interpretation. The above results make it clear that geodemographic segmentation can be used in an effective way to address environmental justice issues.

Without developing a complex statistical model to more thoroughly evaluate specific health and pollution data, at this point, it is difficult to comment conclusively on environmental justice concerns that may arise in Montreal. However based on this segmentation, a useful understanding of relevant economic and sociocultural divisions emerged. Communities that are "positioned" as higher income and with lower numbers of ethnic minorities are seen to the west of the Island and in pockets in Outremont and Westmount. More ethnically groups populate the north-central and eastern parts of Montréal, with income levels being lower in the central region and increasing outwards towards the suburbs on both ends of the island. There is however a very interesting divide in Montreal that permeates traditional notions of social or environmental justice issues, which are largely concerned with racial minority populations. The divide between French speaking and English speaking populations of European descent is the most pronounced division overall. The connection between this cultural division and the economic division is also apparent. The general trend shows an increase in wealth from east to west. Continually, as stated above, the majority of clusters made up of non-European descendants, are largely concentrated throughout the north-central and eastern sections of Montreal. It is therefore reasonable to infer that most "at risk" populations are those located in the dense downtown areas and the eastern sections of Montreal. While this is not new, it is interesting to see how this socioeconomic and cultural spatial division is persisting over time.

There are several ways that this research could be made more robust as the process of using geospatial analysis to support conceptual models for the evaluation of environmental justice and health equity continues. Future endeavors should seek to incorporate a more exhaustive array of indicators, both from the current source (Statistics Canada, 2006 Census) as well as other sources so as to expand the notion of socioeconomic position and environmental health to reflect a diverse range of planning discourse. The incorporation of survey and polling data to represent less tangible and potentially more informative data has great potential. By representing social values, opinions and attitudes, theoretical notions of socioeconomic position can be validated against notions of public engagement at the civic level, potentially assessing the most basic of tenements: participation in the democratic development of healthy, sustainable communities. Furthermore, in the context of public health, environmental justice acts as a viable framework or organizing and mobilizing resources to address basic human needs such as nutrition, sanitation and working environments. Traditionally environmental justice literature has addressed the placement of potentially hazardous land use. Assessment efforts would benefit greatly from the development of indicators that reveal the processes through which urban design and planning decisions are made which allocate access to environmentally beneficial land use, such as neighborhood aesthetics, access to green space and urban waterfronts, recreational opportunities, safe and affordable housing, access to food, and walkability. The aim of including variables such as these rests on the notion that injustice can be found not only be examining how we locate hazardous land use or pollutant producing facilities, but also in the way our neighbourhoods, cities and societies are planned, designed and above all, experienced. In doing so we aim to develop "a more comprehensive approach to environmental policy that often falls outside of the officially defined scope of public health departments and recognize that what constitutes a 'healthy' neighbourhood differs from place to place" (Masuda et al. 2010: 458). Ultimately, progress made in the assessment of environmental justice across Canada will provide a quantitative base upon which the mechanisms of policy development can draw support.

References

Apparicio, P., Leloup, X. & Rivet, P. (2006). *La répartition spatiale des immigrants à Montréal : apport des indices de ségrégation résidentielle.* Montréal : Centre de recherche interuniversitaire de Montréal sur l'immigration, l'intégration et la dynamique urbaine.

Apparidio, P. & Séguin, A.-M. (2006). L'insertion des HLM montréalaises dans le milieu social environnant. *L'espace géographique, 1*, 63-85.

Apelberg, B., Buckley, T., White, R. (2005). Socioeconomic and Racial Disparities in Cancer Risk from Air Toxics in Maryland. *Environmental Health Perspectives. 113*(6), 693-699.

Balkrishnan, T., Gyimah, S. (2003). Spatial residential patterns of selected ethnic groups: significance and policy implications. *Canadian Ethnic Studies Journal*, Spring.

Brulle, R., Pellow, D., (2006). Environmental Justice: Human Health and Environmental Inequalities. *Annual Review of Public Health. 27*, 103-124.

Buzzelli, M., Jerrett, M., Burnett, R., Finklestein, N. (2003). Spatiotemporal Perspectives on Air Pollution and Environmental Justice in Hamilton, Canada, 1985-1996. *Annals of the Association of American Geographers, 93*(3), 557-573.

Galobardes, B., Shaw, M., Lawlor, D., Lynch, J., Smith, G. (2006). Indicators of Socioeconomic Position (Part 1). *Journal of Epidemiology and Community Health, 60*, 7-12.

Galobardes, B., Shaw, M., Lawlor, D., Lynch, J., Smith, G. (2006). Indicators of Socioeconomic Position (Part 2). *Journal of Epidemiology and Community Health, 60*, 95-101.

Hou, F. & Picot, G. (2004). Le visage ethnique des quartiers de Toronto, Montréal et Vancouver. *Tendances sociales canadiennes*, Spring, 9-14.

Longley, P., (2005). Geographical Information Systems: a renaissance of geodemographics for public service delivery. *Progress in Human Geography, 29*, 57-63

Leloux, X. & Apparicio, P. (2010). Montréal, ville plurielle! – Bilan des travaux et perspectives de recherché sur la concentration ethnique. *Nos diverses cités, 7*, 185-194.

Maantay, J. (2002). Mapping Environmental Injustices: Pitfalls and Potential of Geographic Information Systems in Assessing *Environmental Health and Equity. Environmental Health Perspectives, 110*(suppl 2), 161-171.

Masuda, J., Poland, B. and Baxter, J. (2010). Reaching for environmental health justice : Canadian experiences for a comprehensive research, Policy and advocacy agenda in health promotion. *Health Promotion International, 25*(4), 453-465.

Pamuk, A. (2004). Geography of Immigrant Clusters in Global Cities: A Case Study of San Francisco. *International Journal of Urban and Regional Research. 28*(2), 287-307.

Pham, T. T. H., Apparicio, P., Séguin, A.-M, & Gagnon, M. (2010). *Végétation urbaine et injustice environnementale à Montréal : exploration du cas de trois arrondissements centraux à partir de l'imagerie satellitaire à très haute résolution spatiale*. Collection Inédit. Montréal: INRS-UCS.

Rogerson, P. A. (2006). *Statistical Methods for Geography. A Student's Guide*. Thousand Oaks, CA: Sage Publications.

Rast, J. (2006). Environmental Justice and the New Regionalism. *Journal of Planning Education and Research, 25*, 249-263

Taylor, D. 2000. The Rise of the Environmental Justice Paradigm. *American Behavioral Scientist, 43*(4), 508-580

Senecal, G., Hamel, P. J. & Vachon, N. (2005). Forme urbaine, qualité de vie, environnements naturels et construits : éléments de réflexion et test de mesure pour la région métropolitaine de Montréal. *Cahiers de géographie du Québec, 49*(136), 19-43.

Smargassi, A. (2008). *L'influence des inégalités sociales sur les risques associés à la pollution de l'air et la chaleur à Montréal*. Presentation made for the Rencontre francophone internationale sur les inégalités sociales de santé, November 17th, Montréal.

Chapter 5

Observations

Thinking Towards a Tangible and Engaging Health Data Representation

Sven Fuhrmann

Department of Geography, Texas State University, San Marcos, TX

Abstract

The work presented here describes a thought process towards developing a tangible and engaging health data representation that could be used in a community setting to bring "health home". The concept was developed for the second workshop of the ICA working group on Arts and Cartography and involves the placement of differently sized prescription bottles, representing heart-related illnesses for the Island of Montreal, on a 20x20 inch canvas. The underlying simplistic cartographic design and the three-dimensional nature of the representation have the goal to supplement public health communication, to educate the public, to facilitate a dialogue, and to support informed decision-making. As participatory mapping project this approach could support local health benefit efforts and lead to personal transformation.

5.1 Introduction

The general public often makes health decisions at two spatial scales: community-level and individual/family-level. Environmental or other hazardous impacts in a city could result in community health actions, while individual health decisions might evolve around nutrition or exercise related personal preferences (Bucci and Trench, 2008). Many citizens are not health specialists, thus the content and delivery methods need to be easily accessible and well defined. To achieve this goal most health information campaigns utilize advertisements, flyers, newspaper reports, radio and/or TV broadcasts to provide information about a health condition or disease (Logan, 2008). These media are affordable, and reach the general public quickly. However most of the delivery methods are "non-engaging". This particular aspect provided the reasoning to investigate how public health communication could be enhanced with a low cost, accessible, engaging, three-dimensional cartographic representation. The goal was to develop a visual representation that would follow the general cartographic design principles, facilitate health awareness, and potentially influence health attitudes and behavior.

5.2 Developing an Engaging Health Data Representation

Traditionally public health communication is a top-down, one-way communication approach (Rice and Atkin, 2000). Current research indicates that the public is often frustrated with this one-way communication and would prefer in many instances (especially on neighborhood and individual levels) options to ask questions and be involved in a dialogue with scientists, regulatory agencies and decision makers (Stilgoe and Wilsdon, 2009). Spatial representations can effectively facilitate such engaging dialogues with an individual observer and/or in group settings. The core question becomes how to design representations that facilitate dialogues about health. While there are numerous ways to approach this task, the author decided to

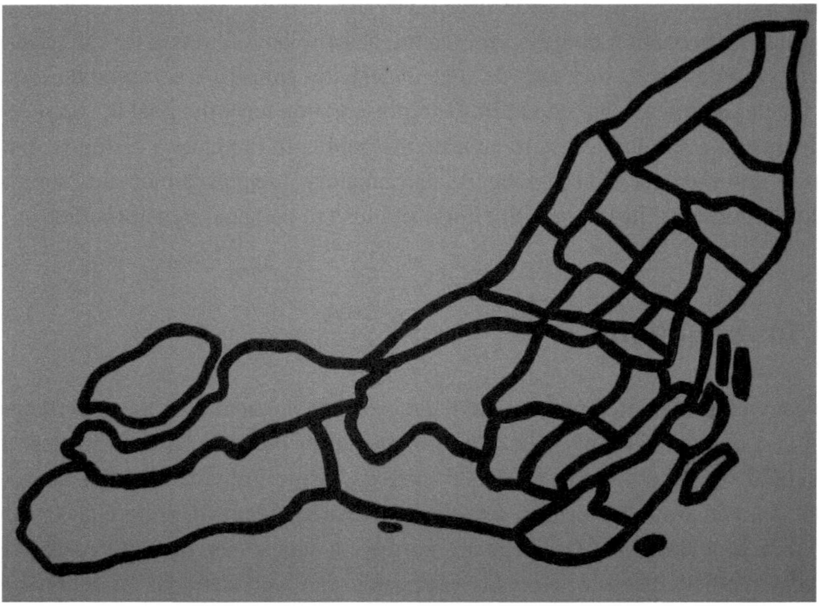

Figure 5.1 Island of Montreal's Twenty-Nine Local Community Service Centers (CLSC), Acryl on Canvas..

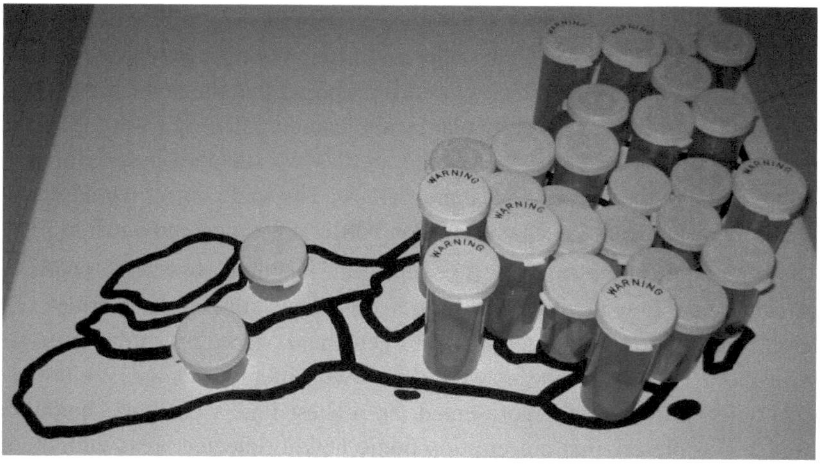

Figure 5.2 Health Data Representation using Prescription Bottles.

address this question by designing a tangible health data display.

The content reported here represents a *thought process* developed for the second workshop of the ICA working group on Arts and Cartography. The workshop was held in September 2010 in Montreal and the organizers provided an extensive Montreal health database to workshop participants who were asked to utilize the datasets for their workshop presentation. Heart related illnesses in twenty-nine local community service centers (health and social service jurisdictions) on the Island of Montreal were the starting point for the health data display design. A 20x20 inch white canvas provided the horizontal and vertical axes to place graphic marks. All twenty-nine local community service centers were projected and traced with black acrylic paint on the canvas. As a result the canvas became literally the "base map"

Figure 5.3 A Hands On Health Conversation with Tangible Objects.

for the health data representation (Figure 5.1).

Twenty nine classic orange and white prescription bottles were used as associative symbolizations (and tangible objects) for representing medicine to be taken for heart related conditions. This particular symbolization will only work for the North American public, since these prescription bottles are standard means for distributing medicine in the United States and Canada. The same object would not work for a European audience, since prescription bottles are almost unknown in Europe.

The heart related illness dataset for twenty nine health regions was classified into three categories: low, medium and high heart-related illness observations. Three different prescription bottle heights were generated according to the three-tiered classification. The plastic bottles were cut at two different lengths. While the full (uncut) height of the bottle represented the highest illness category, bottles with 2/3 and 1/3 of the original prescription bottle height reflected areas with medium and low heart related illnesses. All prescription bottles were placed on the canvas, providing a simple and tangible three-dimensional health data representation (Figure 5.2). While the visual variables "shape" (prescription bottles) and "size" (height of the prescription bottles) provided the initial symbolization; a second variant was also developed by adding a third graphic variable (color hue) to the associative symbolization. The prescription bottle caps were colored in yellow, orange, and red; representing the severeness of the heart conditions in the local health regions (Figure 5.3).

5.3 Conclusions

The goal of this work was to design a three-dimensional, spatial representation that uses associative symbolization, applies basic visual variables and provides a simplistic cartographic design to facilitate public health communication. The display has not been tested and used for health communication. It rather represents a creative process that started with the goal to supplement public health communication, educate the public, facilitate a dialogue, and support making informed health decisions. In collaboration with health and community development experts this initial mapping concept could be further developed into a participatory mapping project in which community members could meet to discuss and learn about health in the city. The simple tangible object approach could also be used (with different symbolizations) for other pressing health topics, such as obesity and smoking. On a larger scale such a participatory mapping project could lead to local action and personal transformation. It would enable informed decisions, provide a hands-on conversation and bring health home.

Acknowledgments

I would like to thank Oak Hill Pharmacy in Evansville, IN and the two anonymous reviewers who encouraged me to revisit and revise this topic.

References

Bucci M, Trench B (2008) Handbook of public communication of science and technology. Routledge, London

Logan R. A. (2008) Health campaign research. In: Bucci M, Trench B (eds.) Handbook of public communication of science and technology. Routledge, London

Rice R. E., Atkin C. K. (2000) Public communication campaigns. Sage Publications, Thousand Oaks

Stilgoe J, Wilsdon J. (2009) The new politics of public engagement with science. In: Holliman R., Whitelegg E., Scanlon E., Smidt S. (eds.) Investigating science communication in the information age. Oxford University Press, Oxford

Chapter 6

Interactive Audiovisual Mapping: BTEX Emissions from NPRI Reporting Facilities in Montreal

Glenn Brauen

Geomatics and Cartographic Research Centre, Carleton University

6.1 Introduction

The *Canadian Environmental Protection Act* (CEPA, 1999) mandates the federal government to assess the potential for adverse effects on human health and the environment from toxic emissions. Under CEPA, Environment Canada maintains a list of priority substances that are likely to be released into the environment in quantities or at concentrations sufficient that they may create “an immediate or long-term harmful effect on the environment or its biological diversity”; that they “may constitute a danger to the environment on which life depends”; or they “may constitute a danger in Canada to human life or health” (ibid., Part 5, Sect. 64). Therefore, Environment Canada (2010c) has established the National Pollutant Release Inventory (NPRI), one focus of which is the collection of annual reports of toxic substance releases by facilities operating in Canada that: a) manufacture, process, or otherwise use one or more of the substances listed under CEPA beyond established thresholds, or b) meet certain facility guidelines such as using an emission stack taller than a specified threshold.[1] These reports, in the form of annual emission totals for each substance from each qualifying facility, are publicly available (ibid.) and include data for a range of compounds. During 2008, facility owners were required to provide reports to the NPRI concerning the use or release of 347 substances or groups of substances, subject to the established reporting guidelines (Environment Canada, 2010a). Included in the list of substances for which reports are required are the volatile organic compounds (VOC) benzene, toluene, ethylbenzene, and the xylene isomers, collectively referred to as BTEX in this chapter and

[1] The NPRI collects data concerning the release of pollutants into air, water, and soils as well as transfer of substances for disposal or recycling (Environment Canada, 2010c). The research reported in this chapter used only the data for air emissions from the Montreal facilities. Although, according to the NPRI data for 2008, the air emissions represent by far the greatest amount of releases for BTEX from these facilities, the use of only the air emissions data from these facilities, and the use only of industrial facility releases, rather than also examining emissions from automobile combustion and other sources, means that the results and mapping to be discussed do not represent total releases or concentrations.

elsewhere. These VOCs are known to pose potential risks for human health, as will be discussed below.

The decision to model BTEX emissions from industrial sites in Montréal for this research was based on the invitation, on the part of the organizers of the *Mapping Environmental Issues in the City* workshop, to work with data concerning cultural, health, and environmental issues in Montreal; the potential for health impacts from these air pollutants; and on the behaviour of these substances as a group. One data set suggested in the workshop announcements was the NPRI annual reports. In particular, Sanford et al (2010) had developed a weighted model of BTEX emissions from facilities in Montréal using the 2008 NPRI air pollutant report emission totals (Environment Canada, 2010c) in conjunction with a model of relative air quality impact for the BTEX constituent compounds. They had then mapped a single interpolated surface of that weighted model for the Montréal area. Although Sanford et al applied weighting computations to reduce spatial variables representing each of the BTEX constituents to a single, visually mapped thematic surface, the availability of the data set and their choice of the BTEX grouping inspired me to revisit the original 2008 NPRI data set and work with the emissions data separately detailing releases of each compound.

Considering the BTEX components as a group suggested that a method of representation that supported simultaneous signification of variables modelling the presence of each compound would be appropriate. Audiovisual cartographic representation, a method developed in earlier research (Brauen, 2006; Brauen and Taylor, 2007, 2008; Caquard et al, 2008; Krygier, 1994; Müller and Scharlach, 2001; Servigne et al, 1999), is one such approach, although not the only method possible.[2]

As will be explained in this chapter, starting with and augmenting the NPRI data for BTEX air emissions from Montréal area facilities during 2008, I developed an interactive audiovisual web map representing dispersed concentrations of the BTEX constituents from the NPRI reporting facilities. In this application, distinguishable sounds are associated with computed airborne concentrations for each of the constituent compounds and the playback of each sound is altered, according to the spatial variation of the estimated pollutant dispersal from the industrial facilities reported in the NPRI, as a user moves the cursor over the map. Although the health impacts of these substances are part of the rationale for examining emitted concentrations from industrial facilities, the application does not attempt to model health impacts from the presence of these substance.

This chapter describes an initial investigation into possibilities for using audiovisual cartography to convey the potential for emissions, reported as released

[2] A review of alternative mapping methods not using combined acoustic and visual representation is beyond the scope of this paper but possibilities include the use of linked graphic variable displays in conjunction with a single map (Monmonier, 1989), or the use of multiple small visual maps (Tufte, 1990).

from point locations, to disperse throughout Montréal. The main motivation for developing this map application and bringing it to the workshop advertised was an interest in trying to understand the Montréal BTEX emissions reported by the NPRI for 2008 and, in particular, to communicate those data, along with contextual information that would aid in comprehending them, in a manner that would be applicable to a broad audience. The focus of the chapter is, therefore, broadly cartographic and describes both the web application as an outcome of a mapping process and the processes by which I worked with and augmented the NPRI data for 2008 BTEX emissions to produce that application.

The remainder of this chapter proceeds as follows. Section 6.2 briefly discusses potential health impacts of the BTEX constituent compounds, and the relevance of industrial BTEX emissions in Montréal. Section 6.3 discusses the data used to develop a map of industrial BTEX emissions for Montréal, the difficulties inherent in estimating dispersion based on the NPRI annual release totals, and design decisions taken to model the emissions, to represent those emissions in an audiovisual web map, and to create an application prototype that is responsive to user interactions with that map. Section 6.4 reflects on some challenges encountered during this research, discusses the audiovsual web map and some responses to it during the workshop, the relationship of this research to other research into audiovisual media design, and remaining challenges and possible future extensions. An appendix at the end of the chapter provides a detailed discussion concerning the configuration of the sound design used to represent airborne BTEX concentrations from industrial facility emissions as reported to the NPRI.

6.2 BTEX: Sources and Potential Impacts

Benzene, toluene, ethylbenzene, and xylenes are all potentially harmful to human health, depending on exposure concentrations and duration. Table 6.1 briefly lists selected human health impacts associated with chronic inhalation of these substances.[3] The BTEX constituent compounds as a group warrant analysis because these compounds are frequently released together, co-occur at waste sites, and the health impacts of these substances may be additive in at least some cases such as their potential neurotoxic actions (Agency for Toxic Substances and Disease Registry, 2004).

The BTEX constituents are common components of petroleum products such as gasoline, and are used in the production of a variety of products including

[3] Other more severe effects may result from acute exposure to higher levels of these substances. These symptoms have not been listed in Table 6.1 but the references cited in the table notes include discussion of these potential impacts.

Table 6.1. Selected potential health effects from chronic inhalation of BTEX.

Compound	Selected potential human health effect (chronic exposure)
benzene[a]	carcinogenic; hematotoxic effects; non-threshold acting.[b]
toluene[c]	central nervous system; upper respiratory system; dizziness; headaches; difficulty with sleep.
ethylbenzene[d]	adverse health effects from low level chronic exposure are unclear (contradictory evidence exists concerning effects on blood); possibly carcinogenic.[e]
xylenes[f]	eye, nose, and throat irritation; gastrointestinal effects; neurological effects (U. S. Environmental Protection Agency, 2003a). Limited studies, mixed occurrence of isomers often with benzene and toluene, have limited the ability for assessment of human health impacts (Environment Canada and Health Canada, 1993b).

[a]Chemical Abstract Services (CAS) No. 71-43-2. Health effects summarized from Environment Canada and Health Canada (1993a) and U. S. Environmental Protection Agency (2003b). [b]"[A] substance for which there is believed to be some chance of adverse effects at any level of exposure" (Environment Canada and Health Canada, 1993a). [c]CAS No. 108-88-3. Health effects summarized from Environment Canada and Health Canada (1992) and U. S. Environmental Protection Agency (2005). [d]CAS No. 100-41-4. Health effects summarized from Health Canada (2007) and U. S. Environmental Protection Agency (1991). [e]"[B]ased on sufficient evidence in experimental animals and inadequate evidence in humans" Health Canada (2007). [f]CAS No. 1330-20-7. Health effects summarized from Environment Canada and Health Canada (1993b) and U. S. Environmental Protection Agency (2003a).

solvents, paints, and pesticides. These substances are released into the environment through combustion, evaporation, or spills from anthropogenic sources and are also produced in the environment from releases not attributable to human activity, but in comparatively lower amounts.[4] Therefore, these compounds are released into the environment in significant quantities in Canada through vehicle emissions; through industrial emissions during production, storage, or transport; through vaporization during and after use; and through accidental leakage and spills (Health Canada, 2007; Environment Canada and Health Canada, 1992, 1993a,b). Despite the levels of release into the external environment, Health Canada (2007) argued that the

[4] Benzene is produced through natural processes at low concentrations (Environment Canada and Health Canada, 1993a). Toluene is produced through the combustion of organic material and is therefore a by-product of forest fires, but Environment Canada and Health Canada (1992) estimated that this source accounted for approximately 4% of total released toluene with the remainder attributable to human activity. Ethylbenzene is most abundantly released through human activity but is also a by-product from combustion of organic material so some releases would be attributable to events such as forest fires (Health Canada, 2007). In addition to releases from the human production, refining, and use of petroleum and coal products, xylenes are "a natural minor component of petroleum" and will sometimes occur in the environment through seepage from or weathering of geological formations, although the amounts of such releases are unknown (Environment Canada and Health Canada, 1993b, 2.1–2.2).

greatest exposure risks, at least for ethylbenzene, may be from indoor air.[5]

Although the BTEX constituents are known to be potentially harmful for human health, depending on exposure levels, other compounds released into the environment around Montréal, from industrial, transportation, or other sources may be of equal or more concern. In developing a ranking of health impacts from pollutants released by Canadian oil refineries, two of which operated in the east end of Montréal during 2008, Gower (2007, 141) argued that the BTEX compounds were among the "air toxics emissions of concern at petroleum refineries across Canada," a group of pollutants in which she included "nickel, PAH [polycyclic aromatic hydrocarbons], BTEX, hydrogen sulphide, sulphuric acid, and vanadium." Gower argued that potential health impacts from particulate matter were greater than for the air toxics listed above (ibid., iii) and that the carcinogenic properties of some of the air toxics may present higher risks than those associated with those constituents of BTEX, other than benzene, for which cancer is not as clearly established as a potential health outcome (ibid., 117). However, she also argued that appropriate caution should be used in interpreting the rankings, and the place of BTEX within them, because of uncertainties concerning potential health impacts compared to variable exposure levels. Whether or not these rankings apply to facilities other than oil refineries and how the emissions from the Montréal area industrial facilities listed in the NPRI compare to transportation or other sources of these compounds would require further study.

Germain et al (2001, iii) argued that motor traffic is the main source of benzene emissions in Montréal. But they noted that industrial emissions predominated at one of the three monitoring stations (Pointe-aux-Trembles) included in their review, at which they determined that detected increases in airborne benzene concentrations were strongly correlated with periods "when the winds were blowing from the west-south-west and western sectors, which is where the industrial plants that reported major atmospheric releases of benzene are located" within 1.5–2 km from the monitoring station (ibid.). Germain et al further argued that benzene monitoring by Environment Canada showed a decline in airborne concentrations between 1989 and 2000 across the island of Montréal, attributed in part to the introduction of gasoline vapour recovery regulations implemented by the municipality. Therefore, although BTEX emissions from transportation in Montréal may generally predominate in comparison to releases from industrial facilities, there is a possibility that releases from some facilities may significantly contribute to spatially variable airborne concentrations in comparison to measurable background levels.

---

[5] In addition to simple off-gassing of these compounds indoors from some of the above products produced using one or more of these substances, BTEX released in the environment from any source can enter soils and groundwater where the substances can dissolve and be transported with the water. Evaporation out of running water is then another path by which these substances can become airborne indoors (Agency for Toxic Substances and Disease Registry, 2004).

Airborne concentrations of the BTEX constituent compounds are then certainly not the only air pollutants that warrant study in Montréal. Neither does the study of industrial releases into the air in isolation from a consideration of vehicle emissions, and potentially other sources, provide a complete understanding of the airborne concentrations of these substances within the city. Rather this chapter is intended to present one means of studying and communicating the dispersion of a group of substances within the city's environment. As discussed above, the NPRI provides estimates of releases concerning a wide range of substances from industrial sources and the approach to mapping them presented here could be applied to other substance groupings included in the annual reports. As will be discussed in Sect. 3.1, this research developed a model of air dispersion suitable for use in creating an interactive audiovisual cartographic representation of air emissions from industrial facilities. If such a model of dispersion from vehicle emissions could be developed, then the audiovisual cartographic representation presented here could possibly be adapted for that model, depending on the run-time processing requirements of the emissions model.

6.3 Model / Interaction / Representation

As will be discussed below in more detail, the data provided in NPRI annual reports require that some assumptions be made concerning what those data represent. A facility report included in the NPRI annual air releases data consists of a minimal facility description (e.g., company name and location) and a set of released substance masses, each entry stating a total mass value for the facility's release of that substance in the specified year through air emissions. Without supplementary data, the only representation of this data that could be constructed would be a simple statement of these total release values associated with the emitter locations. Such a bare bones representation would be rather unsatisfactory because the raison d'être for the NPRI is to assist in the estimation of distribution and impacts of substances, as discussed above, that may get out into the environment and cause harm or increase the risk of harm. Ultimately then this requires that the NPRI reports be used to assess what happens to the emitted substances once released. Where might they go and in what quantities?

Based on this observation, the objective of this mapping process became the development of an audiovisual map that could communicate:

- where and approximately in what concentrations the released BTEX compounds are dispersed throughout Montréal after being emitted; and,
- in an apparent reversal of the dispersion process itself, which, if any, of the NPRI reporting facilities are likely to have contributed to the aggregated estimate of

airborne pollutants from industrial emissions at an arbitrarily selected location.

The remainder of this section provides a design description for the *Montréal BTEX Emission Sources 2008* prototype, developed to address the above objectives. The description is structured using a *model-interactions-representations* (MIR) conceptualization of an interactive map application, as shown in Figure 6.1. This perspective on an interactive map, audiovisual or strictly visual, highlights the components of the application and the relationship between that conception of a map application as a set of active, interacting components and the data sets upon which that application relies for the information it is to represent. The components in this MIR framework correspond to questions that must be answered to determine how the interactive map application under discussion will behave in response to user input and/or actions. Crampton (2002) and Persson et al (2006) each provide classifications of interactive functions that could be considered to suggest possible design approaches for a map application but the MIR framework outlined here is intended as a framework within which any interactive functions chosen can be integrated into a unifying design and assessed to determine whether or not the data necessary to support the desired functions are available.

The *model* is an inventory of the domain knowledge, data, and processing required to create the application. What information is required to create the representations desired and to respond to user actions as desired? Does that information

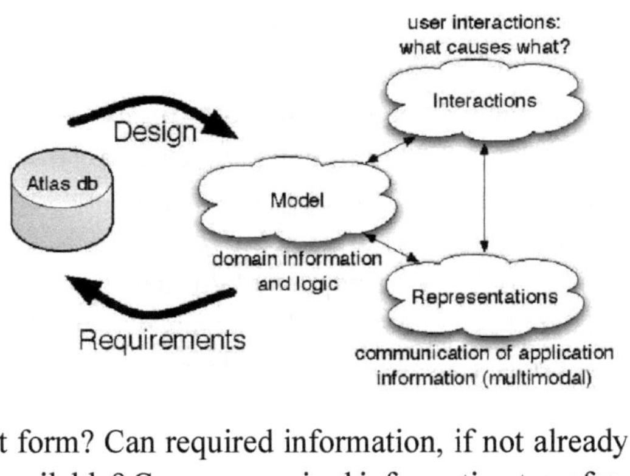

Figure 6.1 Model-interaction-representation conception of an interactive map application.

exist and, if so, where and in what form? Can required information, if not already available, be derived from what is available? Can any required information transformations be completed quickly enough to support user interaction requirements?

Representations is an inventory of the graphic, textual, and auditory information presentations included in the application and is used mainly to generate information requirements for the application but also to rationalize the interaction design of the application. *Interactions* lists user actions to which the application responds and, in conjunction with the representations list, should be examined to ensure that no single action is causing inconsistent responses or too many representational updates, thereby potentially overwhelming users. The set of interactions will also be useful in an analysis of the processing requirements of the application. Does the model contain sufficient information, and in usable and efficient forms, to support the interactions desired?

Designing a web mapping application is often an iterative process; for a map of any complexity almost certainly. Beginning with an initial idea and data set, design efforts assess the feasibility of the original proposal and the completeness and appropriateness of the data available. One result of these efforts will be a preliminary understanding of the domain knowledge needed to support the application and at least a rough proposal of intended interactions and representations. These elements of the design are usually refined through a series of iterations, augmenting and correcting the data available or the modelling algorithms used to complement the existing data and fine-tuning the user interface components of the application. This conception is compatible with cartographic development models such as the *express-test-cycle* methodology proposed by Wood (1994, 16), with the specification of the function, inter-relationships, and data requirements of the interactive map components being part of the *express* phase in that process. The functional division of the interactive map into three components as shown also has similarities to the *model-view-controller* (MVC) design pattern promoted in the Smalltalk-80 programming language (Goldberg and Robson, 1983; Krasner and Pope, 1988) and later object-oriented technologies.[6]

6.3.1 Modelling Air Pollutant Dispersion

Assessment of airborne pollutant dispersion is difficult because of uncertainties concerning: 1) the quantity, timing, and location of releases along with the velocity of the released plume and the height and geometry of the source from which it is released; 2) interactions between the released pollutants and substances already in the environment and the degradation of the pollutants within the environment (Health Canada, 2007; Environment Canada and Health Canada, 1992, 1993a,b); and 3) weather conditions at the time of release including at least wind speed and direction, temperature, and atmospheric pressure (Arya, 1999).

[6] The MVC pattern is a design and implementation pattern whereas the MIR conceptualization of a map application as proposed here is intended to be more of a functional inventory, used to think through what the map application presents to a user; how it produces that representation and using what information; how to address additional data requirements once identified; and what actions on the part of a user are interpreted to cause representational updates and what form those updates will take. The implementation of an interactive map application after an MIR analysis could take many different forms, an MVC design being just one possibility. MVC was also a strictly visual proposal, dealing with only graphics and text, but extends easily to additional modalities with the caveat that, as shown here, a time-based representational media such as sound may create requirements to be able to initiate interaction behaviours, whereas in MVC representations are driven by interactions with no resulting feedback.

The NPRI reported emissions of BTEX from Montréal facilities are, in general, insufficient to accurately understand either the dispersed quantities of each compound or the distribution of those compounds in the environment. Each facility that meets reporting guidelines, must declare the total mass released during a specified year for each substance reported and must include a breakdown showing the percentage of the total released during each quarter of the year.[7] The total mass is reported based on information collected to meet federal, provincial, or municipal regulatory requirements, if the facility is subject to such regulation, or may be estimated using one of a number of methods if the facility is not subject to such regulations (Canada, 2009, 3568). Furthermore, NPRI reports under-represent actual emissions because facilities releasing less than ten tonnes of VOC, or not meeting other minimum criteria for reporting, are not required to provide their data (ibid., 3565). Stack geometry and plume information must be reported only if the facility emits specific substances in amounts greater than prescribed thresholds for each of those substances, and if the facility uses a stack that has a height of 50 metres or more (ibid., 3575).[8] Finally, for each facility a point location is included but even a cursory examination of the locations indicated that some are very rough approximations: one of the Montréal BTEX emitting facilities in the database is located, according to the coordinates provided, in the middle of a four-lane road and another appears to be in the front yard of a suburban home.

6.3.1.1 Generic Dispersion Pattern Computations

Simplifying assumptions have been made to produce a map showing potential compound dispersion based on the NPRI annual reports for 2008. The air dispersion model used for this map is based on a pre-computed set of radial dispersion zones, divided into quadrants by cardinal direction, and centred on a generic point source within a flat, urban area.[9] The U. S. Environmental Protection Agency's AERMOD

[7] Facilities must additionally report a monthly breakdown for Criteria Air Contaminants (CAC) reported to NPRI. The list of CACs includes VOCs "that participate in photochemical reactions, excluding" a list of substances specified in the Canadian Environmental Protection Act (1999, 223–224). The BTEX constituent compounds are not included as CACs and only the quarterly breakdown is provided. The quarterly breakdowns are only available in full downloadable annual data sets from NPRI. The summary data sets and searchable data on their website (Environment Canada, 2010c) only provide the annual total mass for each substance.

[8] Details concerning the stacks are provided as part of the 2008 NPRI data set for only four of the BTEX emitting facilities in Montréal.

[9] This model was adopted after examining the dispersion model used in the HEIDI II health-effects ranking tool and the author acknowledges that the basic structure of the model used here, although applied differently, adheres to the basic approach proposed by McColl et al (2004b). The dispersion factors have been recomputed for this map application.

steady-state air plume modelling software (U. S. Environmental Protection Agency, 2010) was used to compute dispersion factors based on 1996–2000 weather data prepared by the Ontario Ministry of the Environment (OMoE; 2009) for use with AERMOD, and intended for meteorological simulations of urban areas in Eastern Ontario. The data set compiled by the OMoE incorporates hourly surface data from Ottawa, Ontario and upper air data from Maniwaki, Québec.[10]

As shown in Figure 6.2a, the model provides a set of dispersion factors used to convert the NPRI reported annual emission for a source (converted to a continuous mass release rate) into a location-dependent potential pollutant concentration (mass per volume), indexed by the cursor's distance and direction relative to the emission source. The radii of the dispersion zones for a single source range from 500 metres (m) to 30 kilometres (km) with increased zone density within 5 km. For some substances such as particulate matter, limiting the zones of long-term dispersion to 30 km would underestimate the range of impact but with the focus here restricted to VOC and the relatively quick degradation of these substances in the environment, it is not clear that this range limit is entirely unrealistic (Health Canada 2007; Environment Canada and Health Canada 1992, 1993a,b; Gower 2007, 119).

In general, emissions released at lower heights above grade will tend to disperse less widely resulting in higher concentrations close to the facility, although as discussed above plume and meteorological characteristics can influence dispersion. The dispersion zones, as used in the web map application, are pre-computed for each stack height to be used in modelling facility releases. Using AERMOD, a multiplier, used to convert the continuous mass release rates (in kilograms per second, kg/s) to an estimate of airborne concentration (in micrograms per cubic meter, $\mu g/m^3$) contributed by an hypothetical emission source located at the center of the zones, is computed for each quadrant of each radial dispersion zone.11 Because these dispersion zones and their associated multipliers are to be used at run-time in the application to quickly compute additive estimates of pollutant concentrations from all in-range emitters, the conversion values are computed using a flat, generic, urban terrain model (Ontario Ministry of the Environment, 2009). This in effect produces a simple, mobile template of pollutant concentration multipliers that can be used to quickly compute aggregate concentrations for all facilities being modelled with the

[10] The data produced by the OMoE is based on weather data originally obtained from Environment Canada. Environment Canada weather data from stations located in Montréal are also available but not in formats ready for use with AERMOD or its accompanying meteorological pre-processor AERMET. The Eastern Ontario weather data, with the quality control efforts already applied by the OMoE, was deemed an acceptable substitute for Montréal data but more effort determining the impact of this substitution would be warranted.

[11] The AERMOD runs to compute the radial dispersion zone factors all use a source emission rate of 1 g/s and produce concentrations in $\mu g/m^3$. Therefore, the multiplier is in units of $\frac{\mu g/m^3}{g/s}$.

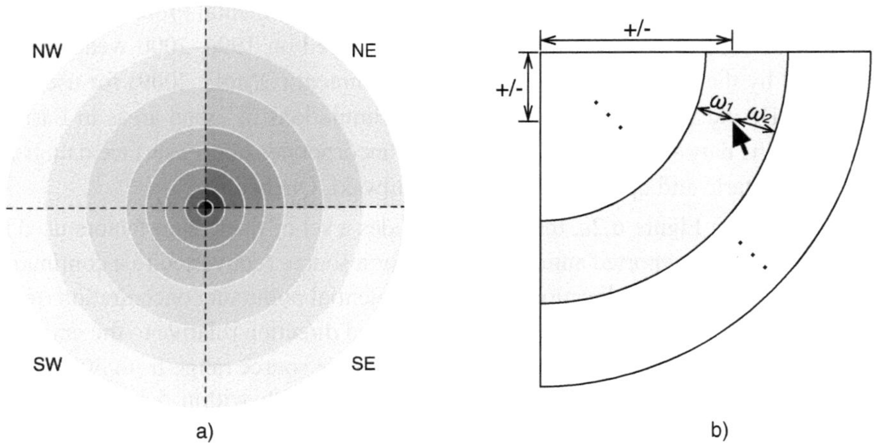

Figure 6.2 Model for emission dispersion showing: a) radial dispersion zones divided into four quadrants; and b) the process of determining the cursor's current quadrant and zone relative to a single source.

same stack height.[12]

The dispersion zones model and associated conversion factors, as a template, allows the web map application to quickly process a user's cursor movements on the map to determine the appropriate dispersion factor to apply to each in-range emission source. As shown in Figure 6.2b, determining within which quadrant the cursor is located, relative to a given source, requires a simple comparison of the source and cursor coordinate values. Once the quadrant and appropriate stack height (defined as a source attribute) are known, the dispersion multiplier, for the source under consideration and the current cursor location, is calculated by interpolating the precomputed dispersion factors associated with each of the radial zones bracketing the cursor (using Inverse Distance Weighting). Then an estimate of potential airborne concentration, c, for each compound (in μg/m³) is computed as

$$c = \sum_{i=1...k} w_i' e_i \qquad \text{(for all in-range sources)} \qquad (1)$$

where w_i' is the interpolated dispersion factor computed for source i (in $\frac{\mu g/m^3}{g/s}$ see footnote 11), and e_i is the continuous release rate derived from the annual release

[12] Following (McColl et al, 2004b), I modelled all stacks using an effective diameter of 0.2 metres, a plume exit velocity of 1 metre per second, and an exit temperature of 50 degrees Celsius. The only variable across dispersion zone computations was the stack height. As stated above, the NPRI facility reports include stack geometry and plume information for a few of the larger stacks, those with heights greater than 50 m. Therefore, in computing dispersion factors for these stacks, NPRI data concerning geometry and plume exit velocity and temperature could have been used, although for each of these larger facilities the NPRI data includes information for multiple stacks so some model of an "average stack" for a facility would still have been needed.

mass for the compound reported for source i (in kg/s). Using this air dispersion model a point location, p, can be used to quickly derive a set of estimates (c_B, c_T, c_E, and c_X) representing the potential airborne concentrations of benzene, toluene, ethylbenzene, and xylenes, respectively, based on the relative direction and distance of p from each of the NPRI facilities reporting BTEX releases.[13]

Within the web map application, all facilities are modelled using a stack height of 15 m, except for those facilities for which the NPRI data includes stack information. For each of these facilities with taller stacks, of which there are four in the Montréal data for 2008, the stack is modelled using a height of 50 m which is, by definition and in practice with this data, a low estimate compared to a computed average of physical stack heights for these facilities. As argued by McColl et al (2004a), the use of a lower stack height may in part account for some amount of fugitive releases in the total air release values reported by each facility. Similarly, for all facilities for which the NPRI data does not include stack and plume information, a stack height of 15 m is intended to account for some fugitive releases in conjunction with the use of a lower main emission stack.

Emissions dispersion is very sensitive to wind direction. The generalized model adopted here incorporates the effects of wind speed and direction as longer term averages into the direction-indexed dispersion factors. Without more details concerning the timing of emissions, this simplified approach to wind effects is justifiable, although the impact of averaging across four cardinal quadrants warrants further study.

6.3.1.2 Background Concentrations: Sanity Comparisons

To sanity check the model-produced concentration estimates, ancillary information was deemed necessary. For this purpose, data from Environment Canada's National Air Pollution Surveillance (NAPS) network of monitoring stations (Environment Canada, 2010b) was used to compute arithmetic mean estimates of background concentrations for each of the BTEX compounds, as shown in Table 6.2.[14] Environment Canada (2008) argued that these measurements do not represent

[13] The run-time computation of potential dispersed concentrations for the Montréal BTEX data set is quite manageable because, of the NPRI reporting facilities in Montréal, only 29 reported releases of one or more of benzene, toluene, ethylbenzene, or xylenes during 2008 and only 25 reported air releases. The map layer for the NPRI reporting facilities is thus pre-filtered so that only these 25 sites need to be considered in response to a user's cursor motions. The approach to the run-time computations would scale to a greater number of reporting facilities if coupled with a Voronoi tessellation of the mapped region based on the locations of reporting facilities that could be used at run-time to only look at those reporting facilities that are potentially within dispersion model range.

[14] The background concentration estimates were computed as the arithmetic means of measured BTEX concentrations during 2008 from two monitoring stations on the island of Montréal. Only

Table 6.2. BTEX background concentration estimates.

	benzene	toluene	ethylbenzene	xylenes
concentration (μg/m³)	1.0	4.2	0.6	2.6

community-wide estimates and are particular to conditions at each monitoring site. These values can then be interpreted only as very rough estimates. But, for lack of any other comparative value with which to assess the concentration estimates from the air dispersion model, these can be and were used to verify that the computed aggregate concentrations at least appear to be of the correct order of magnitude.

After developing the background concentration estimates, they have been kept and used as part of the run-time application for evaluating the computed estimates of dispersion from the emitting facilities. As will be discussed below, these background concentration estimates are used to configure the auditory representation of the pollutant concentrations computed by the dispersion model and thresholds based on these background estimates are used to determine the visibility of each NPRI reporting facility based on whether or not its emissions contribute significantly to the aggregate concentration estimates of at least one of the pollutants at the current cursor location.

6.3.2 Representing Concentrations and Contributing Sites

The web map prototype, *Montréal BTEX Emission Sources 2008*, was designed to:

1. Use sounds to indicate the model computed airborne concentration of each of benzene, toluene, ethylbenzene, or xylenes for the current cursor location.

2. To show, for the current cursor location, those NPRI reporting facilities that contribute significantly[15] to aggregate compound concentrations, as computed by the model, for at least one of benzene, toluene, ethylbenzene, or xylenes.

The web map application is available at http://atlas.gcrc.carleton.ca/ montreal_ btex[16] and is shown in Figure 6.3 highlighting the locations of NPRI reporting

three NAPS monitoring stations in Montréal had complete sets of measurements for 2008: Pointe aux Trembles (50103), Maisonneuve (50115), and Rivière des Prairies (50129). The Pointe aux Trembles station is very close to some of the industrial emitters for which these values are intended to provide a comparative background value. Germain et al (2001) argued that, at least for benzene measurements, this monitoring station is heavily influenced by emissions from the proximate facilities and I thus removed its measurements from these arithmetic means which has, in all cases, lowered the background estimates.

[15] For each of the BTEX constituent compounds, a minimum threshold set at 1/1000 of the background concentration of that compound determined using data from Environment Canada's NAPS stations (see Sect. 3.1.2) is used to filter out insignificant concentration values.

[16] Some components of this web page are designed using Scalable Vector Graphics (W3C, 2010) which are not supported by Microsoft's Internet Explorer. The web page includes notes detailing

facilities in Montréal that released at least one of the BTEX constituent compounds during 2008 and potentially contributed to airborne pollutant levels at the cursor location shown (middle of map — magnified). Based on the air dispersion model described in Sect. 3.1.1, a user's interactions with the map are tracked as a sequence of cursor locations across the surface of the map, each used to compute concentration values c_B, c_T, c_E, and c_X. The results from the sequence of concentration value computations are used to update the audiovisual map representation — both the acoustic representation of airborne concentrations at the cursor location and the visibility of facilities that reported emissions that contribute to the aggregate computed concentration at the cursor location — in a continuing cycle while the cursor remains over the map.

6.3.2.1 Sound Design for Dispersed Concentrations

The use of sound in cartographic design is still relatively rare. Although maps are now predominantly distributed via the World Wide Web, already an important media channel for the commercial distribution of music and other audio material and continuing to gain prominence in these activities, design theory for combining sound and visual elements in maps is not well developed. Krygier (1994) proposed a syntax of acoustic parameters to be used to modify sounds according to the values of spatial variables in conjunction with a visual map. Brauen (2006), Brauen and Taylor (2008), and Caquard et al (2008) provide more current summaries of research into uses of sound as part of audiovisual cartography.

Use of sensory modalities by individuals engaging with the world is contextually based on at least their immediate goals, education, cultural surroundings, and history of personal experiences. When engaging with media, possibilities for engaging sensory modalities are influenced by material capabilities of a medium and the technological, economic, and social understandings of the medium by all those in contact with it, producers and users alike (Gitelman, 2006; Jenkins, 2006). These understandings of a medium will change over time in response to ongoing changes in the cultural context within which the medium operates. Although design theory for audiovisual maps is nascent, there are lessons that can be learned from other disciplines — notably sound design for films and games — within which there has been research and commercial development done to develop theories concerning the production, distribution, and reception of materials that engage sight, hearing, and perhaps other sensory modalities.

Each of film and game sound design have their own histories, discussions of which are beyond the scope of this chapter, but some general motivations for the use of sound emerge from these histories that likely apply to cartographic sound design.

---

with which browsers the application has been tested.

Film sound, in the forms of music, sound effects, and voice, is used to, among other things: increase the willingness of audience members to engage with film (Berg, 1976, 24); establish the reality of the narrative space (Belazs, 1945, 119); establish mood or emotion (Berg, 1976, 197); interact with the visuals to create meaning, beyond what either modality on its own would really signify in a specific context, through what Chion (1994, 5) called *added value*; and to simply deliver information about people, places, contexts, and narrative causality in myriad ways.

Histories of game sound have only begun to appear very recently. Theories of game sound have relied heavily upon lessons learned through research into film

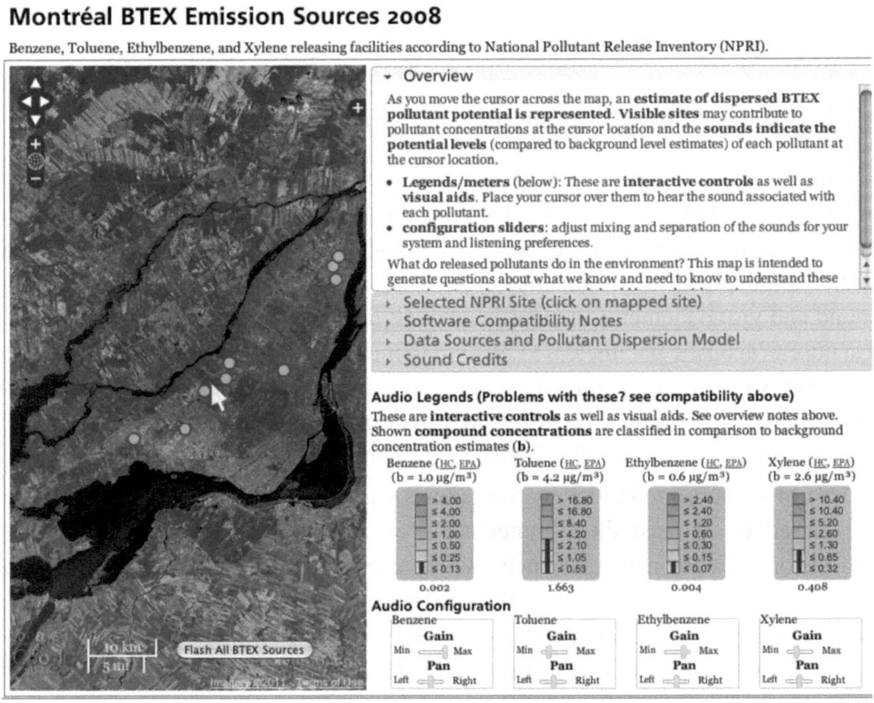

Figure 6.3 Web map application: Montréal BTEX Emission Sources 2008.

sound, and all of the uses of sound listed above could also be applied within games (Collins, 2008, 127–133). But game sound design has also extended that theory, most notably because of the requirement to create sound for the non-determinate narrative trajectories of games (ibid., 125–127); because of the importance of motivating players to continue with a game (Marks, 2001, 188–191); to cue players to impending dangers in game play (Collins, 2008, 133); and to establish rhythms and help a player to physically achieve a target rate of play (ibid., 130).

Within the *Montréal BTEX Emission Sources 2008* application, a distinguishable sound represents the aggregate concentration of each of the BTEX constituent compounds resulting from NPRI reporting facility emissions as computed by the dispersion model for the current cursor location. The gain setting for each sound is determined by classification of the modelled concentration, computed for the latest detected cursor location, according to a set of class boundaries constructed using a logarithmic scale, computed as fractions or multiples of the background concentration estimates (see Sect. 3.1.2).[17] The compound-associated sounds then indicate the computed concentrations relative to background concentration estimates.

For each of the primary auditory indicators used to represent pollutant concentration values, a highly processed or synthesized audio clip has been selected, the main design criteria being: 1) that the sounds blend together well aesthetically with all pairings, or indeed the mix of all sounds together, being reasonably pleasant to listen to; and 2) that each sound be relatively easy to distinguish within the collection so that a user can identify the sounds and learn the associations between each sound and its represented compound. The sound design for the map then combines four primary indicators, each representing one of the BTEX compounds and each audible only when the cursor is over a map location at which the dispersion model indicates that some of the NPRI reporting facilities are contributing a significant amount of the airborne pollutant associated with that sound (relative to its background concentration estimate).

The lowest class for each of these indicators adjusts the gain setting to minimum, rendering the sound inaudible. Although the range boundary values for these classes are at most $\frac{1}{8}$ of the background concentration estimates, they are not necessarily zero. The silencing of the sound perhaps may then suggest erroneously that there is no detectable concentration present, although this is not the intent. Rather this design decision was made to provide some variability in the action of the sound design, allowing individual sounds to drop out of the mix and allow 'sonic space' for those indicators associated with pollutants for which the model indicates elevated concentrations. This is intended to prevent the sounds from becoming overly tedious.

By contrast and as a means of emphasis, a secondary auditory indicator is used to indicate when none of the individual compound-associated indicators are being classified above the lowest class, corresponding to computed aggregate concentrations below $\frac{1}{8}$ of the estimated background concentration. This indicator, using an audio recording of a forested area replete with bird calls but also containing background sounds suggestive of human activity, is used to qualitatively distinguish when the cursor is in a location for which the model estimates that BTEX emissions from facilities that report to the NPRI do not contribute to aggregate concentrations approaching the background estimates.

[17] The ranges are: $c_i \leq \frac{1}{8}b_i$, $\frac{1}{8}b_i < c_i \leq \frac{1}{4}b_i$, $\frac{1}{4}b_i < c_i \leq \frac{1}{2}b_i$, $\frac{1}{2}b_i < c_i \leq b_i$, $b_i < c_i \leq 2b_i$, $2b_i < c_i \leq 4b_i$, $c_i > 4b_i$, where c_i is the concentration estimate for the compound at the cursor location, b_i is the background concentration estimate for the compound, and $i \in \{B,T,E,X\}$.

The representation of pollutant concentrations by highly processed, synthetic sounds is suggestive of the airborne pollutants as outcomes of anthropogenic technological processes. The forest audio counterpoints the processed sounds, suggesting that the areas where this sound is heard are more 'natural', cleaner, and healthier. Although this qualitative distinction was a deliberate design decision, because the model in use here does not factor in all possible sources of BTEX, notably omitting vehicle emissions, and also does not include data concerning other potentially hazardous pollutants, that distinction is very possibly overstated by this choice. Because the lowest class for each of the compound-associated sounds silences the audio, the use of an additional sound to signal the condition when all of those sounds are silenced is, from a data signification perspective, redundant. But as suggested above, and as will be discussed below in more detail, that perspective does not account for the creation of meaning through connotation. Therefore, although the sound is not necessarily required to reinforce the fact that no other sound can be heard, that is not why it was added and that clearly is not all it seems to indicate to users.

Also shown in Figure 6.3, dynamic audiovisual legend components are shown, in the lower right of the application window, that act as both learning aids and user interface controls. First, as learning aids, the dynamic legends, one associated with each compound, display the airborne concentration estimates computed based on the current cursor location and show the values being classified according to the class ranges discussed above. These visual elements reinforce the sound design by graphically explaining the classification scheme in use much as a legend on a static map would. In addition, the dynamic audiovisual legends are updated in synchronization with the sound design updates to assist a user in understanding how the sound changes for variable values that are differently classified. Second, as active user interface controls, each dynamic legend enables a user to isolate the individual sound associated with one compound by positioning the cursor over the legend, thereby directly controlling the current value of the associated concentration and initiating the sounding of the acoustic representation for that concentration value. As is shown in Figure 6.4, when a dynamic legend is engaged as a direct controller, the position of the cursor over the classification grid determines the value for the associated variable and the associated sound is adjusted according to the representation of that class.[18] Sounds for all other variables are silenced, partly because the cursor is not over the map and thus defines no current value for them but also to enable individual sounds to be isolated for a user to learn them.

[18] Although the compound concentrations are represented using classification in this map, variables could also be represented without classification, for example, by linearly interpolating the value within a defined *min ... max* range. The dynamic audiovisual legends, when used as direct controllers, then use the cursor position over the legend scale as the set point for interpolating a value for the associated variable.

The gain and pan sliders shown in the bottom right of Figure 6.3 allow a user to adjust the sound design to separate sounds using left-right panning controls or adjust relative gain for each audio indicator as necessary to accommodate peculiarities of their computer, the attached speakers or headphones, and their own listening capabilities and preferences. Additionally, these can be used to silence one or more of the acoustically represented variables, if not of concern during a particular map use session.

A more detailed explanation of the sound design for the *Montréal BTEX Emission Sources 2008* prototype is provided in the appendix.

6.3.2.2 Visual Design for Contributing Sites

If the dispersion model computations for a specific NPRI reporting facility, $w_i' e_i$ for facility i from equation (1) on p. 11, yield a compound concentration deemed insignificant compared to the cut-off threshold (footnote 15), then that contribution is not included in the aggregate concentration for that compound at the current cursor location. If a specific NPRI reporting facility is deemed to not contribute to any of the BTEX concentrations for the current cursor location, c_B, c_T, c_E, or c_X, then that facility will be rendered invisible on the map until the cursor moves to a location for which the dispersion model computations determine the facility does contribute to the aggregate concentration of at least one of the compounds. Facility visibility on the map then indicates qualitatively whether or not each facility is contributing to the concurrently represented BTEX concentrations. Therefore, the visibility of a facility signifies that it is contributing to emitted BTEX concentrations at the current cursor location, according to the dispersion model. The radius of the largest dispersion zone used in this model is 30 km and that distance is, therefore, a hard limit on the representation of facility emissions within the application. If the

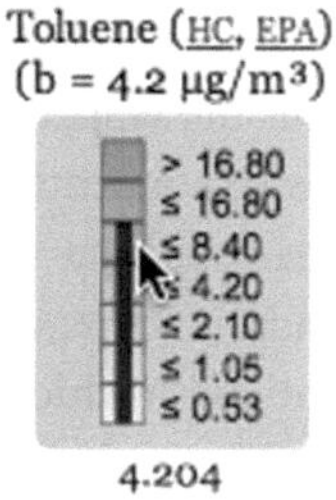

Figure 6.4 Toluene audio meter being used as sound controller. The scale is computed based on the estimate of background concentration, b.

cursor moves more than 30 km from a site, regardless of the reported emissions from that site, that site will become invisible.

The effects of the pollutant dispersion model visibility rules can be seen in the map included in Figure 6.3, with only nine of the possible facilities being visible as a result of the shown cursor placement. Another example of the effects of these visibility rules is shown in Figure 6.5. Both Figure 6.5a and Figure 6.5b show the same area of Montréal with each showing eight visible facility locations. In each figure, the cursor is positioned in the proximity of two sites that appear only in that image and thus emit relatively lower amounts of BTEX compared to the sites that appear in both figures. Despite the cursor being well within the maximum range of 30 km for the estimated concentration computations, the two sites close to the cursor in Figure 6.5a fade to invisibility once the cursor has moved toward the upper right of the figure as shown in Figure 6.5b. This change in visibility indicates that these facilities probably do not emit sufficient amounts of any one of the BTEX compounds to be significant concerns for contributing to the aggregate concentrations at the new cursor location.

6.3.3 Responding to User Interactions

The sound design reacts as a user moves the cursor over the map, fading up highly processed audio when the dispersion model suggests the cursor location is subject to airborne BTEX emissions from NPRI reporting facilities and fading up a mix of environmental sounds, notably highlighting bird calls, when that model suggests that airborne BTEX emission concentrations are low.

The sound design responds to user actions in concert with updates to the visibility of the NPRI sites, also based on estimates of airborne pollutant concentrations from the dispersion model as discussed earlier. The overall audiovisual design uses sounds to indicate estimates of airborne pollutant concentrations, also supported by the visual updating of the dynamic audiovisual legends, and uses the visibility of the NPRI facility markers to indicate which of the known emission sources may be contributing to the sounding concentrations. Auditory updates are implemented as fades, gradual transitions toward an intended update value, both to avoid harsh acoustic transitions and to de-emphasize the suggestion of clear boundaries. Similarly the visibility of the represented NPRI reporting facilities also fade in and out over time as the dispersion model computations gauge whether or not each facility contributes to concentrations at the cursor location. Although the dispersion model and the existing data are determinate, uncertainties in both suggest that drawing clear boundaries around the emission zone for any single source could exaggerate the authority with which this map should represent this data. Although site visibility is determinate — a specific site will be shown or not based on nothing but the current location, *if the cursor remains in that location long enough for the fade to be fully accomplished* — the interaction of a user moving the cursor across the map with the temporal lag in visibility introduced by the fade timers will tend to blur the range of visibility of a site.

The visually and acoustically fading representation of this data was chosen as a deliberate strategy to suggest the characteristics of uncertainty and instability with

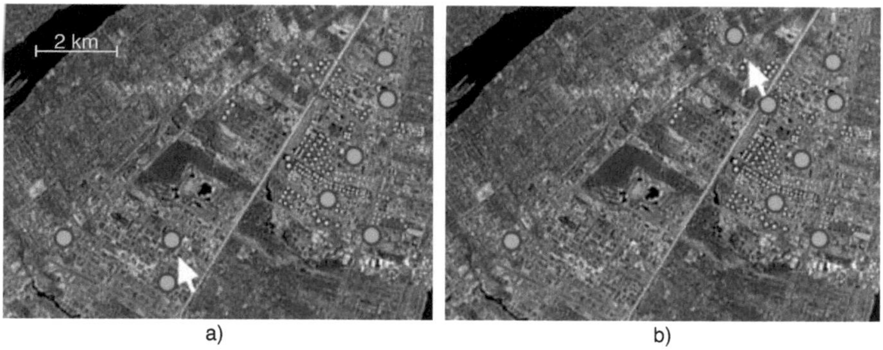

Figure 6.5 NPRI reporting facility visibility as a function of cursor location.

respect to the actual circumstances of release and the dispersion of these pollutants from NPRI reporting facilities in Montréal and with respect to our awareness and knowledge of these emissions and their effects. The map's audiovisual design is finally a communicative strategy intended to highlight important information about these emissions and is an interim outcome of a process of raising questions concerning what more would need to be studied to fully grasp where and when these pollutants are in the environment and in what concentrations.

By limiting the visual representation of emitting sites to a point marker, rather than for example a scaled proportional symbol based on the quantity of a site's releases, the map remains uncluttered and is able to effectively communicate what information the NPRI does clearly provide: the identities of the corporate emitters, their approximate location, and the reported release quantities for 2008.

The design of the audiovisual representation of dispersed concentrations and contributing sites, and how these respond to cursor movements, were intended to be the predominant elements of the user interaction design for the *Montréal BTEX Emission Sources 2008* application. However, additional information concerning the emitting facilities, and the areas in which they are are located, can also be obtained using the application's map interface and information panels. If a user wishes to obtain more information concerning the NPRI reporting facilities displayed on the map, they can: 1) click on a site marker, causing information for that facility, from the NPRI reports, to be displayed in the right hand panel of the map application as shown in Figure 6.6; 2) zoom in on the location using the map navigation controls (as in Figure 6.5); 3) switch to a base map layer that includes street labels to better locate visible facilities; or 4) click on the *Flash all BTEX sources* button (positioned near the bottom of the map in Figure 6.3) to briefly show all of the mapped NPRI reporting facilities before fading them back to invisibility. The first three functions allow a user to obtain more information about any single facility or to better locate a facility or group of co-located facilities within the city.

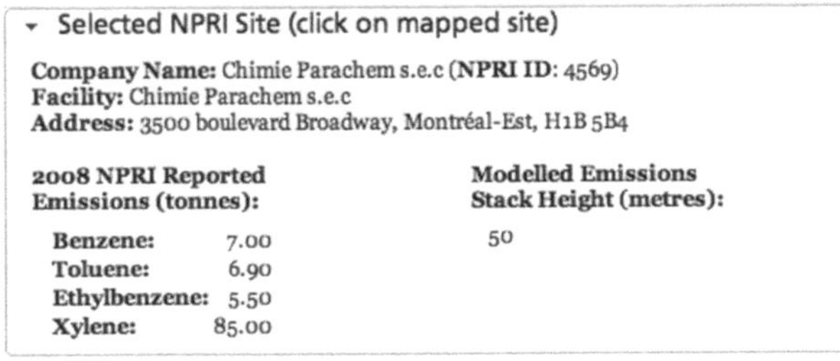

Figure 6.6 Information display for an NPRI reporting facility.

The last function, flashing all sources, provides a user with a means to see the overall distribution of BTEX emitting facilities within the city. The audiovisual design is such that a synoptic overview of the overall pattern is normally not provided and one user commented that this was initially an unwelcome characteristic of the map. In addition, because facility visibility on the map is a function of the quantity of BTEX emitted, some facilities only appear if the cursor location is very close to the facility or, as is the case with a few of the facilities in the current data, placed directly on the facility. Therefore, allowing a user another means to obtain such a synoptic view of the facilities using a function such as this, or by pairing this type of map with other visualizations, seems quite reasonable.

6.4 Discussion

This paper has outlined research conducted to develop an audiovisual web map prototype concerned with industrial releases of BTEX in Montréal. The objective of this research was to design a web map that would communicate the locations of BTEX emitting facilities and, for each, an indication of the contribution that site makes to airborne pollutant concentrations according to an air dispersion model run as part of the map. Conversely, the map is also intended to identify for an arbitrary location, selected by positioning the cursor over the map, those emitting facilities that are probably contributing to the airborne concentrations of at least one of benzene, toluene, ethylbenzene, and xylenes.

The resulting web map application is, I believe, an intuitive and perhaps deceptively straightforward seeming representation of potential BTEX emissions. Driven by user interactions, particularly cursor motions over its surface, the map makes only those emitting facilities visible that the underlying dispersion model indicates are probably contributing to the aggregated concentrations from industrial emissions at the cursor location. Auditory representation, coupled with the visual assistance provided by a set of audiovisual dynamic legends, is used to classify the concentrations resulting from NPRI reported emissions of each compound at the cursor location. Participants at the workshop at which the first prototype of this map was presented seemed to quickly understand the map representation and the feedback was positive, although only limited informal testing has been conducted in which individuals manipulated the map themselves.

The main points of discussion concerning the map's representation of BTEX concentrations at the workshop were:

- An audiovisual approach to mapping information was readily accepted and at least one participant expressed that they thought the design adopted for this map, fading sites and sounds, was very well suited to mapping NPRI emissions data with its attendant uncertainties.
- The counterpointing of the highly processed sounds representing compound concentrations with a naturalistic environmental recording, prominently featuring bird calls, was questioned both for the obvious stylistic difference between that sound and the rest but also for the symbolism of positioning "Edenic purity" as *the* alternative to the presence of airborne contaminants. Sarah Kanouse argued that this symbolic representation implied that the removal of all airborne contaminants, an unlikely scenario, would be the only way to improve environmental health and that this representation could then potentially foreclose a broader discussion of alternatives. However, at least one workshop participant also argued that they thought the secondary audio selection was effective in communicating that the emissions do not seem to be a concern in some areas.
- Whether or not there was an intended mood implied by the selection of the audio samples used to represent BTEX emission concentrations, with at least one participant suggesting that they sounded "scary." The sounds, as discussed above, were selected to suggest anthropogenic activity in relation to the emissions. Some of them, such as the highly processed synthetic sounds of helicopter rotors associated with ethylbenzene,[19] because of industrial and potentially militaristic associations possible with such a sound and because of pop-culture references with which that sound could be related, allow for a range of characterizations that could include 'ominous.'

A remarkable consideration in the discussion of the use of the environmental audio to reinforce the condition in which the dispersion model classifies all of the BTEX concentrations as being in the lowest class range is that, during the discussions, the decision to reinforce this with a sound was never questioned. Sound has to date been rather uncommon in maps, typically used as an ambient background or organized spatially by the map rather than used as a representational component of the map, despite earlier work by Krygier (1994) concerning sound in interactive maps and by Rice and Rice (2001) concerning the functions of music with respect to

[19] This particular sound is edited from a clip called 'crello04.wav' posted to the freesound project, an audio sharing online community at freesound.org, by Tim Kahn (freesound user 'Corsica S'). Credits for all of the audio samples used in the *Montréal BTEX Emission Sources 2008* prototype are listed in the application web page.

animated maps. Harrower (2007), with respect to interactive maps, and Vetere and Howard (2000) and Kalyuga et al (2004), with respect to educational multimedia, have suggested that redundant information presented using sound can reduce learning effectiveness, especially when narration is used to duplicate information available visually such as text or images. Although the subject audio in this discussion was not narration, I did note that there was no suggestion that the sound was unnecessary and instead the discussion immediately focused on what that particular representation meant. I interpret this workshop discussion as suggesting that Chion's (1994) arguments concerning audiovisual *added value* can and do transfer to audiovisual mapping.[20]

Whether or not the sounds were intended to engender any specific interpretation of the subject matter or instill a particular mood as part of the representation of the data is similarly noteworthy. The status of visual maps as outcome of processes codifying certain knowledges and interpretations of the world in preference to other perspectives (Harley, 1989; Pickles, 2004), and as condensing propositions and thereby imposing material effects on the world (Kitchin et al, 2009; Krygier and Wood, 2009; Wood and Fels, 1986; Wood, 1992) are topics that are well covered by the literature. Nevertheless, the literature concerning cartography does not often grapple with emotions or mood directly although recently Craine and Aitken (2009) have called for this type of engagement. Considering the strong associations, in for example film and game sound theory as discussed in this chapter, between affect and sound, more widespread adoption of audiovisual methods within cartography are certain to require that the emotional consequences of design choices receive more attention. It would be disingenuous to pretend that the sounds used in the prototype were selected only because as a set they were internally distinguishable and that I did not think about how they would or could be interpreted. But it is also worth noting that other participants in the workshop argued that they did not interpret those sounds that way. Apart from considering that sounds will have interpretations then, cartographers will need to anticipate a range of interpretations when using sound. Similarly, with respect to the choice of the recording of a forested area as a counterpoint to the compound concentration sounds, although I agree with some of the concerns raised about that choice of audio, I have left it in place because I think that it highlights a potentially productive discussion concerning the nascent topic of sound design for audiovisual cartography.

The discussion of the map in this chapter has been structured around a framework conception of an interactive map as composed of inter-related components — model-interactions-representations. This provides a useful division allowing

[20] Chion (1994, 5) argued that sound in combination with film visuals may guide a viewer to interpretations of the combined audiovisuals that in retrospect seem to have been apparent all along but, without sound, may not have materialized.

the logic underlying a map application's features to be discussed by separating discussions of the spatial data and modelling logic from discussions of how those elements will be represented and how a user is expected to interact with and come to understand the model presented by the map. More widespread use of audio in cartography, and potentially other sensory modalities (Taylor, 2003; Taylor and Pyne, 2010), will add complexity to the representation and interaction designs adopted and, at least until some standard models for these emerge, will require that design and explanatory frameworks be adopted that support the communication and sharing of methods.

The audio design on its own, partly because of the rarity of sound design discussions in relation to cartography, stands out as an element that warrants the separation of representation discussions from other aspects of the design. A sound design for an audiovisual map, even one such as this that used only variable-driven gain adjustments, presents a designer with significant flexibility, and thus decisions that need to be made, including selection of audio; whether or not to normalize represented data; whether or not to use classification and, if so, what class value ranges to use; timing decisions related to parameter fades; and final mixing adjustments of the entire design. Decisions related to the data such as those concerning classification and normalization apply equally to visual and auditory representations but there seem to be fewer obvious models to which a map maker can turn when considering those decisions related to how a mix of variables will sound. The appendix provides a more detailed discussion of the application sound design.

Since the workshop, I have spent more time reassessing the data issues underlying the map, reworking details of the dispersion model and searching for and using better weather data than I had at the time. There is more that could be done. Even now, as discussed above, the weather data still substitutes a generic Eastern Ontario urban area for Montréal (a displacement of approximately 200 km), and is based on averages derived from 1996–2000 rather than data for 2008 that would match the NPRI emissions data. The use of the NAPS data to compute a single background concentration estimate for each compound applied to the entire mapped region is simplistic. The dispersion model is also based on a flat terrain model and so ignores the topographical features of Montréal. Although the weather data in the dispersion model could be improved, it is less clear how well a more complex model of background concentration levels would work as part of the representation of NPRI facility emissions.[21] And while it seems intuitive that a more accurate terrain model would render better dispersion estimates, it is not clear how such a computation would work in an interactive map application. The dispersion model adopted in the

[21] Developing a spatially variable model of background concentrations would cause the classification of computed facility emission in relation to that background (i.e., as a function of the difference between background concentrations and emitted concentrations from facilities) to be variable for multiple reasons and therefore more difficult to communicate.

prototype, because based on a generic template of emissions from a generic facility, can be applied at run-time in response to user actions, whereas a more complex model may not be workable as part of an interactive interface.

Importantly, it is not clear that effort to make the weather and terrain data more accurate would result in better estimates of dispersed pollutants because it is unclear by how much the NPRI reports underestimate actual emissions.[22] An unknown number of facilities are not required to report emissions. The timing of releases is unknown with, at best, quarterly emission totals reported for BTEX. Facility details such as location are obviously approximate and stack configurations through which reported emission are released are often not reported.

Equally importantly, the data addressed by this current prototype is only a tiny subset of industrial emissions that could be studied and mapped. As with visual thematic maps, the audiovisual representation methods described here will only concurrently support a limited number of variables and map usability may suffer if a greater number are added. Beyond a relatively small number, designing a collection of sounds that are distinguishable while listenable becomes difficult. There are other logical groupings of chemicals within the NPRI data to which I expect this approach would be applicable. To apply these methods to larger numbers of compounds would, again as in the case of visual mapping, require that the number of concurrently mapped variables be reduced, possibly by combining emissions data for multiple substances using weightings developed according to some criteria such as health impacts.

The research described here has not directly incorporated estimates of health impacts into the web map prototype, preferring instead to classify the computed concentrations according to estimates of background levels. This choice has meant that this prototype models whether or not the facilities as a group, and specific facilities with respect to specific locations, are increasing the airborne pollutant concentrations but goes no further to, for instance, suggest possible effects of such an increase. Providing some context to understand the data is important and with the focus selected for this application, the Environment Canada NAPS data does provide some context in allowing modelled concentrations to be compared to monitored concentrations. Other choices would be possible. For example, it would be possible to gauge the modelled concentrations by comparing them to standards like Ambient Air Quality Criteria such as are produced by the Ontario Ministry of the Environment (2008), thereby allowing some suggestion of impacts

[22] Deslauriers and Niemi (2005) argued that NPRI reported emissions for 2003, when compared to the then latest national inventory of criteria air contaminants (CAC), underestimated across the major categories of CAC with the NPRI reported emissions for VOC accounting for only 10% of those estimated by the national inventory. Deslauriers and Niemi further argued that new processes being put in place for reporting emissions to NPRI would probably increase both the number of reporting facilities and the percentage coverage of emissions.

to be introduced in the representation. Such a view to alternative mappings and representations would be consistent with one of the motivations behind the adoption and development of interactive geographic visualization methods: the ability to develop multiple views of data including comparisons between alternative data sets but also between alternative representations of the same data (MacEachren, 1994). Therefore, the research reported here should be understood as one approach to studying geographical data as part of a group of such methods, applicable to a wide range of types of data, rather than as simply a discussion of a standalone audiovisual mapping prototype.

6.5 Conclusions

The research reported here has applied interactive audiovisual cartographic techniques to study reported releases of BTEX from Montréal-area facilities during 2008. In producing this web map application, data has been found and combined into a model of aggregated airborne concentration estimates from industrial emissions, as reported to the NPRI, for these compounds across the region.

Audiovisual web mapping has not been extensively researched by cartographers and is an approach that warrants further research concerning design possibilities, application to various subject domains, and communicative effects. The Montréal BTEX Emission Sources 2008 web map prototype resulting from this research was well received when presented at the workshop. Based on the reception received, the time and effort required to create and refine the map to its current state (see chapter appendix), and the results obtained in attempting to apply this mapping technique to a subject domain to which it had not previously been applied, I think this project was successful. The fading sites and sounds design applied to the NPRI annual emissions data represents that data but attempts to not overstate the authority with which the dispersion can be accurately predicted.

As discussed above, the emissions modelled by this prototype represent only one type of source for the emissions of BTEX, let alone the other pollutants that are in the environment. As such, this prototype should be understand to be a demonstration of an approach to studying and communicating geographical data in a form applicable to a broad audience, as compared to the specialist audience of modelers capable of applying air dispersion simulation software directly, rather than as an attempt to model total pollutant concentrations. Further study would be required, possibly using and extending the approach discussed here, to compare the emissions from other types of sources, and of other compounds to determine an overall understanding of either the actual dispersion of pollutants in the environment or any potential impacts of those pollutants.

An analytic approach based on permanent, independent monitoring of airborne emissions, such as that provided by Environment Canada's NAPS stations, is useful to augment the NPRI data. Such an approach on its own would be insufficient to trace concentrations back to sources but, in augmenting the source reported emissions, can provide a useful gauge of whether or not the modelled emissions seem reasonable. The NAPS network of monitoring stations is currently sparse and is not intended to provide estimates of community-wide concentrations so that data must be used with care.

Acknowledgements

This research was supported in part by the Cybercartography and the New Economy project, that was in turn funded by the Social Sciences and Humanities Research Council (SSHRC) of Canada under an Initiative on the New Economy (INE) Collaborative Research Grant. Dr. D. R. Fraser Taylor was the principal investigator. In addition, Glenn Brauen's research has been funded in part through a SSHRC Canada Graduate Scholarship. Thanks to two anonymous reviewers for providing constructive criticisms of an earlier draft that have helped me to improve this contribution. I am responsible for remaining errors and omissions.

Appendix

The description provided here is intended to elaborate on the sound design overview presented in Sect. 3.2.1. From the brief overview provided there, it is clear this design requires that:

1. Each sound be developed and maintained as an independent element of a composite overall implementation allowing each sound to be played, stopped, or processed according to the representational design relating that sound with mapped data.
2. Each sound be controllable and modifiable while a user is working with the map. The indeterminate, non-linear interactions of a user working with a map precludes the possibility of preloading a correctly processed auditory representation. The sound design must be able to react to a user's actions with minimal delay if the user is to interpret modifications in the sound as related to their interactions with the audiovisual map.

These characteristics stress that the map as distributed must retain digital media characteristics as outlined by Manovich (2001, 27–48). The distribution of the sound design as part of the audiovisual map over the World Wide Web places all of the elements of that design simultaneously both in a cultural context, as sounds intended to be heard and understood as such, and in a technological context, as numerical representations that can be transmitted across a network and are subject to the behaviours of the network and computer systems sending, receiving, and forwarding them.[23]

The sound design for the *Montréal BTEX Emission Sources 2008* prototype has been designed and implemented using an audiovisual web mapping sound subsystem (Brauen and Taylor, 2007, 2008). Figure 6.7 provides an abstract overview of the components used to create the auditory representation for one spatial variable, referred to collectively as an *audio indicator*, in the sound subsystem for an hypothetical application. An auditory indicator is created using an *audio source*, a prerecorded digital audio file in the examples discussed in this chapter and shown as circles in Figure 6.7.[24] The audio source for an auditory representation is connected to the computer's audio output through a series of *audio effect processing components*, shown as squares in Figure 6.7, each of which typically manages a single parameter of that source's sound output (e.g., gain, pan, mute, or play/stop settings). The source and effects processing set is controlled by the current value of an associated spatial variable, with a *variable monitor*, shown in Figure 6.7 as a circular sector, passing detected value updates for the variable through a *classifier object*, the output of which is then passed to all of the audio effect processing components to modify the audio being output by the indicator.

These control updates are designated using dashed lines in Figure 6.7 that indicate the run-time reconfiguration of the effect processing components — adjusting gain settings, mute controls, or other digital signal processing effects — in response to variable value updates causing changes in the class selection for the variable. The

[23] In addition to the design assuming that a user's computer and web browser are capable of decoding and rendering the map and associated multimedia elements as intended by the map's author, use of the map may also be dependent on the network connecting a map user's computer to the system hosting and distributing those sound definitions being able to transmit those sounds in a timely manner. Sampled audio requires a significant amount of information to accurately reproduce a sound and thus requires a network that is capable of transmitting information quickly and consistently. Sampled audio may be used as a stream whereby the sound reproduction begins at a user's computer before the entire sound definition is transmitted, if indeed it is possible to think of the stream as being a finite sound definition (e.g., Internet radio streams are effectively unbounded). Conversely, sampled audio may be used as a clip, whereby it is transmitted and stored on a user's computer before being played, with the understanding that playing the file multiple times only requires that it be transmitted once and, after that initial transmission, the use of the sound is no longer subject to the possibility of network delays.

[24] In the most general case, an audio indicator is created using one or more *audio sources*, each a prerecorded digital audio file or an instrument in a synthesized composition, and each source connected to the computer's audio output through a series of audio processing components. In the audio design for the map prototype discussed in this chapter, all of the auditory representations of variables are created using only a single source with its associated effects chain. Therefore, the discussion here presents only this more simplified abstraction.

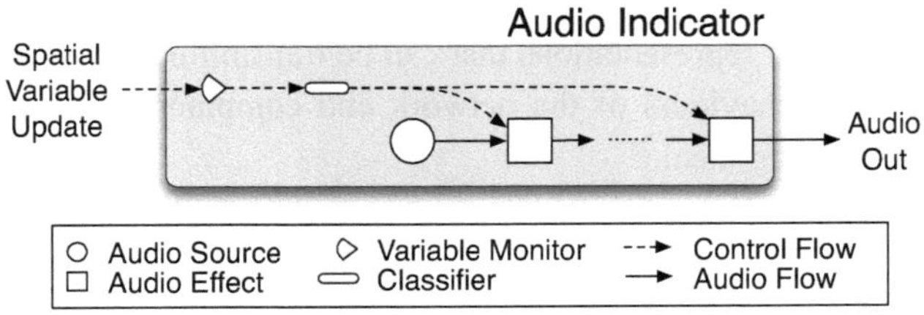

Figure 6.7 Audio indicator components.

solid lines in Figure 6.7 show the audio data path running sequentially through the set of effect processing components, the flow of which is broken by an ellipsis to show that the audio indicator design supports a variable number of effects processing components that can be configured as needed by a given design.

An audio design for an audiovisual map in the most general case then includes, but is not restricted solely to, a set of audio indicators, each associated with a different variable. The user interface components of the map application provide event handling routines that detect cases such as the cursor being moved over certain regions of the map or the dynamic legends. These event handlers in turn are written to issue function calls to the sound subsystem application programming interface (API). The sound subsystem API provides interfaces that can be called to update the variable monitor created for an audio indicator to a new value, thereby triggering sound subsystem processing, as depicted in Figure 6.7, leading to an altered sound output for that indicator.

Figure 6.8 shows a schematic of the sound design for the *Montréal BTEX Emission Sources 2008* prototype. For each of the compounds, the design includes a variable monitor that is updated using the dispersion model concentration variables discussed in Sect. 3.1.1, a source consisting of a short audio clip played in a loop, and a single attached gain control that is managed according to a preconfigured classification scheme. When an update cycle occurs for a concentration variable, the current value is classified resulting in the selection of a class index that is subsequently used as an index into a configured array of gain control values (shown below each gain control). The actual gain control values need not be discussed here except to notice that for class 0 each gain control adjusts the gain setting to 0, causing the sound to fade to silence. Higher class indices correspond to higher gain settings thus the control scales the loudness of the audio clip using the metaphor *louder means more*. The classification objects are shared between the audio indicator design and the dynamic audiovisual legend design discussed in Sect. 3.2.1. The number of classes is thus the same as shown in Figure 6.3 and Figure 6.4 and the class range boundaries are the same as discussed in footnote 15 on p. 13.

In addition to the primary audio indicators for the emission concentrations, an additional secondary audio indicator has been included in the sound design. Shown as *all concentrations low* in Figure 6.8, this indicator uses a summation of the class indices from the primary compound indicators as its monitored variable. Again the only audio control configured for this indicator is a classifying gain control, designed to make the audio stream audible only when all of the primary indicators are set to class 0 (the only case where the sum will be zero, the condition specified for a value to fall within class 0 according to the configuration of this indicator's classification object) and are thus silent.

The configured gain for class 0 of the *all concentrations low* source audio is essentially arbitrary, as are all configured design options, but was selected to adjust the loudness of this sound within the overall sound design. Some elements in this audio recording seem very loud compared to the other audio clips and scaling the configured gain settings enables a designer to adjust for this. In effect, this is equivalent to the selection of appropriate saturation or hue settings for a sequence of fills when designing a colour scheme for a map. Experimentation is required to create a configuration that works well, integrating an ensemble of individual indicators and dynamically controlling the output of each indicator across the range of its acoustic variations.

The sound design, possibly because of the amount of white noise included in some of the source audio clips, sometimes exhibits impulse artefacts (snaps and crackles) during gain fades. The cartographic sound subsystem in use allows for arbitrary effects to be created and connected but I have not yet implemented filtering effects that may be able to correct these crackles.

Based on my previous work to develop the cartographic sound subsystem used to create the prototype discussed here, the actual effort to implement the first version of this map for the workshop was approximately five (long) days, after dispersion model alternatives had been studied and the one described had been selected. Additional time has been spent since the workshop refining the data used in the dispersion model, re-editing the audio samples used, and clarifying the information panels associated with the map.

The sound subsystem described here is a programmer's toolkit. No end-user programming application has yet been developed to allow a broader user group to take advantage of this technology. The results produced by others attempting to use this software could vary. An early version of this software was released as free and open source software (FOSS) as part of the *Nunaliit Cybercartographic Atlas Toolkit* (Geomatics and Cartographic Research Centre, 2006). Although many updates have been done to the sound subsystem software since that toolkit was released, and the intent is still that this be made available as FOSS, efforts have not been made to produce a new release of this as a stand-alone tool. Anyone interested in using this for their own projects is encouraged to contact the author.

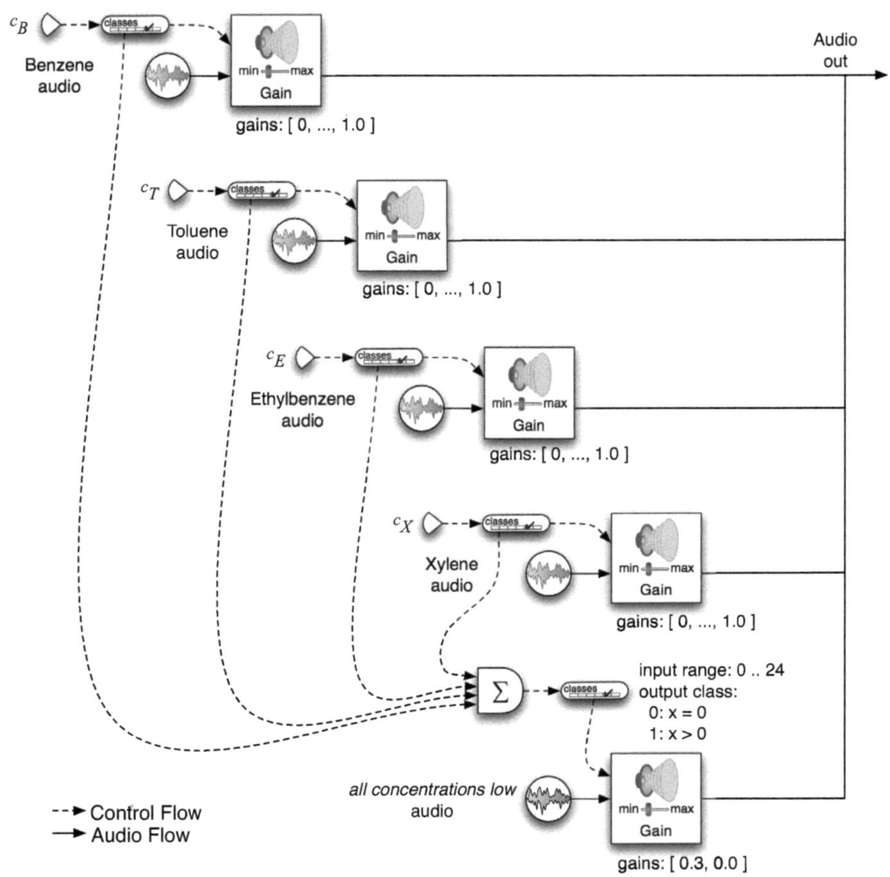

Figure 6.8 Sound design for *Montréal BTEX Emission Sources 2008* application.

References

Agency for Toxic Substances and Disease Registry (2004) Interaction profile for benzene, toluene, ethylbenzene, and xylenes (BTEX). U.S. Department of Health and Human Services. Available at http://www.atsdr.cdc.gov/ interactionprofiles/ip05.html. Accessed February 2011

Arya SP (1999) Air Pollution Meteorology and Dispersion. Oxford University Press, New York; Oxford

Belazs B (1945) Theory of the film: Sound. In: Weis E, Belton J (eds) Film Sound: Theory and Practice, Columbia University Press, New York, pp 116–125

Berg CM (1976) An Investigation of the Motives for and Realization of Music to Accompany the American Silent Film, 1896-1927. Arno Press, New York

Brauen G (2006) Designing interactive sound maps using Scalable Vector Graphics. Cartographica 41(1):59–71

Brauen G, Taylor DRF (2007) A cybercartographic framework for audible mapping. Geomatica 61(2):127–136, Canadian National Commission of the International Cartographic Association National Report on Cartography

Brauen G, Taylor DRF (2008) Linked audio representation in cybercartography: Guidance from animated and interactive cartography for using sound. Revista Brasileira de Cartografia 60(3):223–242

Canada (2009) Notice with respect to substances in the National Pollutant Release Inventory for 2009. Canada Gazette 143(49):3550–3579

Canadian Environmental Protection Act (1999) Bill C-33. Assented to 14th September 1999. Ottawa: Public Works and Government Services Canada

Caquard S, Brauen G, Wright B, Jasen P (2008) Designing sound in cybercartography: From structured cinematic narratives to unpredictable sound/image interactions. International Journal of Geographical Information Science 22(11):1219– 1245

Chion M (1994) Audio-Vision: Sound on Screen. Columbia University Press, New York

Collins K (2008) Game Sound: An Introduction to the History, Theory, and Practice of Video Game Music and Sound Design. The MIT Press, Cambridge, MA; London

Craine J, Aitken SC (2009) The emotional life of maps and other visual geographies. In: Dodge et al (2009), chap 9, pp 149–166

Crampton JW (2002) Interactivity types in geographic visualization. Cartography and Geographic Information Systems 29(2):85–98

Deslauriers M, Niemi D (2005) Compilation of annual emission inventories in canada for criteria air contaminants. Paper read at 14th International Emission Inventory Conference. Las Vegas, Nevada. Available at: http://www.epa.gov/ ttn/chief/conference/ei14/session6/ deslauries.pdf. Accessed March 2011

Dodge M, Kitchin R, Perkins C (eds) (2009) Rethinking Maps: New Frontiers in Cartographic Theory. Routledge, London; New York

Environment Canada (2008) National air pollution surveillance (NAPS) network: Annual data summary for 2005 and 2006. Report En49-2/7-39. Available at http://www.ec.gc.ca/Publications/default.asp?lang=En&xml= 8928E8A8-0C6F-4B1B-8597-AB3E0F31F4A5. Accessed December 2010

Environment Canada (2010a) Listing of national pollutant release inventory substances for 2008. Available at http://www.ec.gc.ca/inrp-npri/default. asp?lang=En&n=C3FF94C3-1. Accessed February 2011

Environment Canada (2010b) National air pollution surveillance (NAPS) network. Available at http://www.ec.gc.ca/rnspa-naps/. Accessed December 2010

Environment Canada (2010c) National pollutant release inventory (NPRI). Available at http://www.ec.gc.ca/inrp-npri/. Accessed December 2010

Environment Canada, Health Canada (1992) First priority substances list (PSL1) assessments: Toluene. Report En40-215/4E. Available at http://www.hc-sc. gc.ca/ewh-semt/pubs/contaminants/psl1-lsp1/toulene/index-eng. php. Accessed December 2010

Environment Canada, Health Canada (1993a) Benzene. Report En40-215/11-E. Available at http://www.hc-sc.gc.ca/ewh-semt/pubs/contaminants/ psl1-lsp1/benzene/index-eng.php. Accessed December 2010

Environment Canada, Health Canada (1993b) First priority substances list (PSL1) assessments: Xylenes. Report En40-215/22E. Available at http://www.hc-sc. gc.ca/ewh-semt/pubs/contaminants/psl1-lsp1/xylenes/xylenes_3eng.php. Accessed December 2010

Geomatics and Cartographic Research Centre (2006) Nunaliit: Cybercartographic atlas framework [software]. Published by Geomatics and Cartographic Research Centre, Carleton University. Available at http://nunaliit.org. Accessed October 2010.

Germain A, Rousseau J, Dann T (2001) Issues related to benzene in eastern Montreal. Tech. Rep. EN40-644/2001E-IN, Environment Canada, Montreal, Quebec

Gitelman L (2006) Always Already New: Media, History and the Data of Culture. MIT Press, Cambridge, MA

Goldberg A, Robson D (1983) Smalltalk-80: the language and its implementation. Addison-Wesley, Reading, Mass.

Gower S (2007) A computer-based decision tool for prioritizing the reduction of airborne chemical emissions from Canadian oil refineries using estimated health impacts. Ph.d. thesis, University of Waterloo, Waterloo, Ontario

Harley JB (1989) Deconstructing the map. Cartographica 26(2):1–20

Harrower M (2007) The cognitive limits of animated maps. Cartographica 42(4):349–357

Health Canada (2007) Ethylbenzene and health. Report H128-1/07-496-3E. Available at http://www.hc-sc.gc.ca/ewh-semt/pubs/contaminants/ ethylbenzene-eng.php. Accessed December 2010

Jenkins H (2006) Convergence Culture: Where Old and New Media Collide. New York University Press, New York; London

Kalyuga S, Chandler P, Sweller J (2004) When redundant on-screen text in multimedia technical instruction can interfere with learning. Human Factors 46(3):567–581

Kitchin R, Perkins C, Dodge M (2009) Thinking about maps. In: Dodge et al (2009), chap 1, pp 1–25

Krasner GE, Pope ST (1988) A cookbook for using the model-view controller user interface paradigm in Smalltalk-80. Journal of Object Oriented Programming 1:26–49

Krygier J, Wood D (2009) Ce n'est pas le monde (This is not the world). In: Dodge et al (2009), chap 11, pp 189–219

Krygier JB (1994) Sound and geographic visualization. In: MacEachren and Taylor (1994), chap 8, pp 149–166

MacEachren A (1994) Visualization in modern cartography. In: MacEachren and Taylor (1994), chap 1, pp 1–12

MacEachren A, Taylor DRF (eds) (1994) Visualization in Modern Cartography. Pergamon, New York

Manovich L (2001) The language of new media. MIT Press, Cambridge, MA

Marks A (2001) The Complete Guide to Game Audio: For Composers, Musicians, Sound Designers, and Game Developers. CMP Books, Lawrence, KA

McColl S, Gower S, Hicks J, Shortreed J, Craig L (2004a) Development of a health effects-based priority ranking system for air emissions reductions from oil refineries in Canada. Waterloo University Institute for Risk Research. Available at: http://www.irr-neram.ca/research/eh/eh.html

McColl S, Gower S, Hicks J, Shortreed J, Craig L (2004b) Health effects indicators decision index (heidi ii) [excel workbook]. Waterloo University Institute for Risk Research. Available at: http://www.irr-neram.ca/research/eh/eh.html

Monmonier M (1989) Geographic brushing: Enhancing exploratory analysis of the scatterplot matrix. Geographical Analysis 21(1):81–84

M¨uller JC, Scharlach H (2001) Noise abatement planning: Using animated maps and sound to visualise traffic flows and noise pollution. In: Proceedings of the 20th International Cartographic Conference, International Cartographic Association, vol I, pp 375–385

Ontario Ministry of the Environment (2008) Ontario's ambient air quality criteria. Tech. Rep. PIBS# 6570e, Standards Development Branch, available at http: //www.ene.gov.on.ca/environment/en/resources/STD01_076461.html. Accessed march 2011

Ontario Ministry of the Environment (2009) Air dispersion modelling guideline for Ontario, version 2.0. Tech. Rep. PIBs # 5165e02, Toronto, available at http://www.ene.gov.on.ca/envision/air/regulations/localquality. htm. Accessed December 2010

Persson D, Gartner G, Buchroithner M (2006) Towards a typology of interactivity functions for visual map exploration. In: Stefanakis E, Peterson MP, Armenakis C, Delis V (eds) Geographic Hypermedia: Concepts and Systems, Springer, Berlin; Heidelberg, chap 15, pp 275–292

Pickles J (2004) A History of Spaces: Cartographic reason, mapping and the geocoded world. Routledge, London

Rice K, Rice S (2001) Scoring and scripting music for animated maps. Paper read at the 21st Annual Conference of the North American Cartographic Information Society (NACIS)

Interactive Audiovisual Mapping: Montréal BTEX Emissions

Sanford A, Ramasamy A, Marsan-Paquin JF, Thompson U (2010) Air quality estimation based on industrial release: BTEX toxicity index, course project for Geographic Information Systems for Environmental Impact Assessment (ENVS663), Concordia University, Winter 2010

Servigne S, Laurini R, Kang MA, Li KJ (1999) First specifications of an information system for urban soundscape. In: IEEE International Conference on Multimedia Computing and Systems (ICMCS'99), vol II, pp 262–266

Taylor DRF (2003) The concept of cybercartography. In: Peterson MP (ed) Maps and the Internet, Elsevier Science, Oxford, chap 26, pp 405–420

Taylor DRF, Pyne S (2010) The history and development of the theory and practice of cybercartography. International Journal of Digital Earth 3(1):2–15

Tufte ER (1990) Envisioning Information. Graphics Press, Cheshire, Connecticut

U S Environmental Protection Agency (1991) Integrated risk information system: Ethylbenzene. Available at http://www.epa.gov/iris/subst/0051.htm. Accessed December 2010

U S Environmental Protection Agency (2003b) Integrated risk information system: Benzene. Available at http://www.epa.gov/iris/subst/0276.htm. Accessed December 2010

U S Environmental Protection Agency (2003a) Integrated risk information system: Xylenes. Available at http://www.epa.gov/iris/subst/0270.htm. Accessed December 2010

U S Environmental Protection Agency (2005) Integrated risk information system: Toluene. Available at http://www.epa.gov/iris/subst/0118.htm. Accessed December 2010

U S Environmental Protection Agency (2010) Support center for regulatory atmospheric modeling (SCRAM). Available at http://www.epa.gov/ttn/scram/. Accessed December 2010

Vetere F, Howard S (2000) Prior knowledge and redundant multimedia. 2000 IEEE International Conference on Multimedia and Expo 2000 2:605–608, DOI 10. 1109/ICME.2000.871436

W3C (2010) Scalable Vector Graphics (SVG) 1.1 Specification. 2nd edn, world Wide Web Consortium. Available at: http://www.w3.org/TR/SVG11/. Accessed October 2010

Wood D (1992) The Power of Maps. The Guildford Press, New York

Wood D, Fels J (1986) Designs on signs / myth and meaning in maps. Cartographica 23(3):54–103

Wood M (1994) Visualization in historical context. In: MacEachren and Taylor (1994), chap 2, pp 13–26

Chapter 7

The City as MetroMap: Envisioning Montreal

William Cartwright

School of Mathematical and Geospatial Sciences, RMIT University, Australia

Abstract

MetroMaps provide representations of urban transportation systems that 'work' by simplifying the geography of a city and the network of services that travel through, over or under the urban extent. They allow commuters to identify a suitable route between places and to determine appropriate interchanges if needed. These maps simplify 'messy' multiple information resources, by providing (geo)graphic tools to understand a system and to navigate from place to place.

Urban transportation systems link place-to-place and the MetroMap is used to illustrate these links. MetroMaps 'slice' their way through cities, linking neighbourhoods that may vary socially, economically and politically. MetroMap slices can be employed to 'travel' through a city, and, if appropriate artefacts are appended at each place, the user can compare one place to another during a mental trip in the city. In this way users can gain an appreciation of how the nature of a city changes as we move through it.

Here, MetroMap refers to the concept of employing the metro map metaphor for affording more efficient means of delivering geolocated graphic information. The metro map of Montreal will be used as the foundation for provisioning geo-placed artefacts to illustrate the environment of linked places in the city. It will be used as a framework upon which 'found' (virtual) objects can be hung. These artefacts will be sourced via the Web (WebCams) and Web 2.0 (for example images from *Flickr*® and videos from *YouTube*). Artefacts will be sourced according to Elkingham's (1997) 'Triple Bottom Line", which is used to measure an organization's impact and sustainability at both a local and a global scale - the *environment*, *social & cultural*, and *economic*. To this will be added an additional item – *governance*, the '+1' extension to Elkingham's Triple Bottom Line, proposed by GS@RMIT (Global Sustainability at RMIT University (GS@RMIT, 2010).

The resultant image of the city will be delivered as a Web 2.0 application, providing, where possible a recent or current (using real-time images from WebCams) images of the city and delivered as MetroMap slices of the city. This concept was developed as part of a demonstration prototype to provide an environmental 'picture' of Montreal, by feeding information from public Web resources and private electronic 'chatter' about the city (here, via *Twitter* ®) to provide place-specific information about the city.

7.1 Introduction

The reality of contemporary life in a city is that sensors continually monitor and report on things like air quality (from Web-linked monitors), traffic flows (via stationary cameras), pedestrian movements (using CCTV), traffic light operation (through the use of reports and system management commands sent via SIM-enabled street fixtures), shopping transactions (with RFID tags), etc. The city is monitored, measured, controlled and managed using formalised wired and wireless devices.

In a recent special report in *The Economist* (November 6th 2010) this growing phenomena was used to illustrate the growth of 'Smart Cities' that are using formal sensor systems to assist in better measuring and management of urban environments. One article: "Living on a platform" (pp. 9-10), noted the rise in purpose-built systems in newly constructed cities like Masdar in Abu Dhabi and Songdo City (near Seoul) in South Korea and 'retro-fitted' installations in cities like Singapore and Beijing. These so-called 'smart-city' projects are managed with data provided from connected control systems.

The Economist report noted that such a connected world was proposed in the early 1990s by David Gelernter in his book Mirror Worlds (1992). Mirror Worlds is a model of institutions or processes linking to personal or external information. It was described as "a past/present/future system, a miniature software world, complete in every (useful) detail, leaving a historical record behind" (Gelernter, 1994, p. 32).

However, the use of less formal systems to glean information about a city can also be used as information resources. As individuals have ready access to consumer electronics devices like smart phones they increasingly use them to record and communicate more personal, and placed, information about the environment where they live or through which they pass. Information can also be published using blogs, shared via *Facebook* and other social media and distributed using *Twitter*®. This geo-located information is stored in 'the cloud', transmitted via wired or wireless Internet means and made available for general access by the 'wired community' or as more limited access resources for more personal, corporate or governance applications. The information is immediate, focused, pertinent and, most importantly, in the context of this chapter, placed. Users are able to interact with (many) computers (through wired or wireless means) and construct more personal 'pictures' of a neighbourhood or a city.

Similarly, Gelernter proposed a means of interacting with computers like this that he named lifestreams (Gelernter, 1992, 1994; Steinberg, 1997) - an interface that would hold and display data linked to a personal chronological database/personal information 'grab bag'. Gelernter also saw the potential of universities using a space-time interface offering " a new way to organise books, pictures, videos and objects of all sorts that inhabit the trackless depths of the information ocean" (*op cit.*, p. 32).

He described his envisaged system:

"It would work roughly like this. You sit down at a computer and find a world map on the screen (this system caters mainly to the historians, journalists, social scientists, students, wonks and spies whose interests tend to centre on particular times and places). You steer the computer (using a mouse) to your time and place of interest. You now see on your screen a map that is zoomed into your region as it appeared during your time period. ... You see blips on the screen, localised on the map as far as possible. Each blip represents a library holding or archive or government or museum holding somewhere in the world that bears on this place and time.

*Now type a description of your particular interests to the computer. Blips that marks out relevant objects turn red, irrelevant objects turn blue, and maybes (*sic*) are various shades of purple. Mouse on a blip to learn more. Human experts and knowledge seekers appear as blips too, if they are willing. The system allows you to send them mail. What better way to find the world's only expert on early 10th Century Breton goat herding, or leave a query about a CIA operative who taught at Yale in 1953?"*

Now, the availability of numerous devices that can feed information, in an automatic or ad hoc manner, may go some way in providing the data sources for developing a real system like Gelernter proposed.

However, with the explosion of the number of devices that are able to collect placed information – for example nearly 270 million smartphones were sold in 2010 (*The Economist*, 2010, p. 18) – means that there needs to be a means for accessing and understanding this information. Here, it is suggested, that the use of metaphors, and, in particular, the MetroMap metaphor, needs to be explored as a tool for delivering information about a city. This information can be 'harvested from consumer electronics-captured data, from Web-delivered reports, newspapers and blogs, from feeds from Web cams and RSS feeds and from the content of electronic messages sent via communication systems like *Twitter* ®.

Thus, this chapter reports on the development of a demonstration prototype Web-delivered information resource that was designed around the MetroMap metaphor and provides information to the user that is place-determined and delivered using Web 2.0 services that provided Web feeds of data that illustrates the Triple bottom line +1 elements of a city. The city used for the development of the package, as a proof-of-concept demonstration prototype, was Montreal, Canada.

To describe the ideas behind the design of the package and the development of the demonstration prototype the chapter first begins by talking about a different map that is possible via the Web. It then outlines the advantages of using metaphors to describe complex information. Then the MetroMap metaphor is addressed and a general outline of its roots made. After this, a brief overview of gaining an understanding of the environmental profile of Montreal during the International Cartographic Association Working Group on Art and Cartography's Workshop "Mapping" Environmental Issues in the City: Arts and Cartographic Cross Perspective", held in Montreal between September 8th to 10th 2010 is provided. This is followed by a description of the demonstration prototype package. Finally, areas of potential further development and research are addressed.

7.2 Visualising the City with Contemporary Mapping

Mapping has at its core the requirement to accurately show spatial phenomena. It must be scientifically and mathematically accurate – and show information in its correct geographical position – and artistically and aesthetically accurate – designed to maximize map use and information communication and understanding. The 'stuff' that comprises the discipline is measurement and depiction. Designers and producers of map products are concerned with whereness - something that can be formulated and depicted in quantitative terms, and whatness - dealing with qualitative information. The whyness element of mapping is a combination of a user's knowledge about the subject being depicted and the cartographic package developer's skill in choosing the appropriate data and designing the most effective portrayal medium.

The depiction of 'somewhere in space/time' depends on a number of elements - the choice of the method of graphic portrayal, the attributes of the information that have to be depicted, the influences on how the nature of the data and its location may alter, the catalysts for change that bring-about the final location in space/time for particular data elements, and the rules and conventions that need to be employed. These rules and conventions are those that relate to the type of data being depicted, the viewing preferences of the user and the specific demands of the visualization method and delivery method being employed.

The real value of contemporary mapping – from paper map to Web-delivered products generated from collaboratively assembled information, although providing timely and accurate products, is sometimes compromised by the very nature of the graphics themselves or the delivery mechanism employed. There now exists plethora of other information delivery devices that can be appended, linked, incorporated within or used in conjunction with currently available cartographic products. However, in many instances, the real power of 'appended' geo-referenced information resources is not fully exploited to enhance and make more relevant 'mainstream' mapping products. Maps (in their widest sense) can be complemented by what is being offered to localize and personlaise what is being depicted on the map.

In designing conventional maps, it is usually the intention of the designer to impart specific and relevant information to users, who absorbs this information and thereby extend their existing view of reality. Appending annotations to mapping products may enhance and expand the user's reality about a place. Annotated mapping product designers need to look beyond the traditional methods and seek to provide knowledge to users, as well as data and information. Technology now provides the means of publishing information-rich annotated maps, to better impart knowledge about place.

7.3 'Lived Experience'

Accessing information from 'lived experience' enhance information being provided and represented through maps and complementary annotations. Using the medium of print, authority is given to the information telling due to the fact that the author is absent and therefore the artistry and the quality of the subsequent representation of place gives credibility to the data presented. Electronic delivery gains this authority and credibility by the ability to provision a user with accurate, timely and comprehensive information. Content is important, but the context is coloured by the credence of the 'teller'.

It is this richness, it is argued, that annotated mapping applications can provide. By tapping into the lived experience of those who reside or work in a place, providing selections of appropriate reports and articles and observations about that place, and supplementing these with information 'feeds', in real-time or near real-time, better representations of place could be provided.

It is with these thoughts in mind that a brief exploration of annotated maps was undertaken. It was done to investigate how a better understanding of a city – in this case the City of Montreal, Canada – might be had if users were able to exploit the availability of information feeds that would annotate maps. The environment of

Montreal was the topic chosen for the development of a demonstration prototype and the 'Triple Bottom Line' was employed as indicator elements that would link the information nodes together.

7.4 Demonstration prototype basics

The concept of the demonstration prototype was guided by three key elements:

- A model for gauging the health of the environment of a city. The model selected was the 'Triple bottom line' Elkington (1997).
- A map metaphor that exploited the potential of providing geo-located information effectively and allowing for information about the city's environment to be easily read and comprehended. Information graphics provided a solution and the MetroMap metaphor in particular was chosen to facilitate this.
- Map annotations that would be digitally appended to the metro map, providing insight into the environment of the city.

Triple Bottom Line

Triple bottom line measures an organisation's impact and sustainability – locally and globally. One element of this project was to ascertain if this measure could be extended to a city. Measurement is done in reference to three elements:

- Environment: human and man-made;
- Social & cultural: indicators, codes, legislation, initiatives and ethics; and
- Economic: Genuine Progress Index (GPI), full cost accounting, product stewardship and supply chain economics

The human environment refers to both the impact that humans have on an environment – as they live in or move through and environment. How we live has a direct impact on where we live. Man-made environment is that infrastructure, buildings, communications and transportation systems and how we change an environment impacts upon the very nature of where we live.

Social and cultural indicators are those things that 'signals' how ordered (or disordered) a society actually is. Whether codes or legislation are in fact required sends a direct signal about whether a society can self-regulate, or if it needs laws imposed upon it to ensure a workable society. 'Signposts' of community initiatives – for example, active neighbourhoods, self-supportive projects, urban renewal or refurbishment, proper maintenance of personal, shared and public spaces and buildings – provide indicators about the 'health' of a society.

Economic accountability and management are reflected in the availability quality of community resources, accessibility to all citizens and the delivery of services in a timely, planned manner, with due consideration of economic constraints and 'value for money' for taxpayers and funding organisations.

As well as these three indicators, Global Sustainability at RMIT University proposed a '+1': Governance. The elements to be measured as part of this indicator are done at both global and local levels and look at voluntary and mandatory elements of what can be considered to be 'Good Governance'.

These indicators can be monitored in a city and reported via 'standard' reports, usually supported by maps. This gives an official 'report card' about the city, but, in many instances, ignores the real human elements of a city. To gain a true picture about well how a city is 'travelling' when measured according to the 'Triple bottom line +1' additional information needs to be collected to enhance and personalise knowledge about a city.

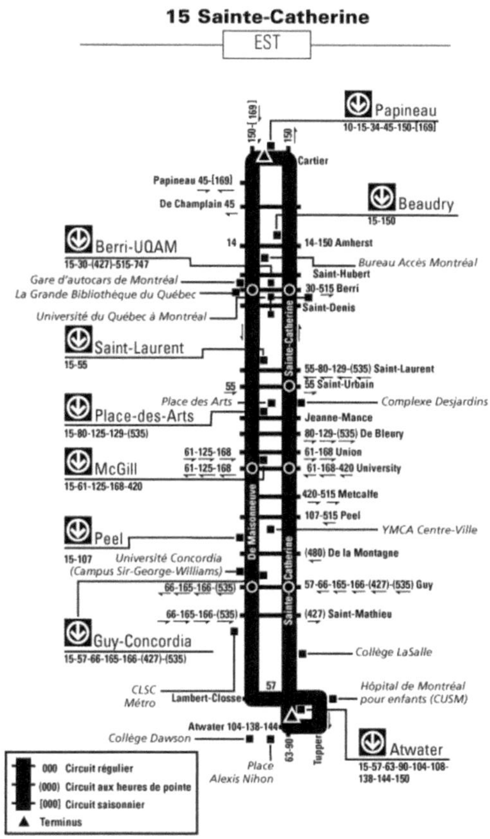

Figure 7.1 Montreal STM bus route No. 15.
http://www.stm.info/English/bus/plan_lig/A-PL15.htm

What is now addressed is how these indicators might relate to Montreal? A pictorial overview of this follows.

During the Workshop *"Mapping" Environmental Issues in the City: Arts and Cartographic Cross Perspective"*, held at Concodia University in Montreal, Canada between September 8th and September 10th 2010 participants had the opportunity to explore a 'slice' of Montreal. This was done by taking a guided bus ride with the International Cartographic Association Working Group on Art and Cartography's co-Chair and Workshop conveynor, Dr Sebastien Caquard. This entailed workshop participants riding a local bus (No. 15 Sainte-Catherine), which crosses the city from East to West, providing the opportunity to gain a quick appreciation of the general socio-economic profile of the city. The bus route map is shown in Figure 7.1.

The trip began at Concordia University (Guy-Concordia bus stop) (Figure 7.2a), then travelled through the downtown area (Figure 7.2b). The group then walked through a pedestrian area (Figure 7.2c) and ended at the Berri-UQAM metro station (Figure 7.2d).

Figure 7.2a Start at Concordia University.

Figure 7.2b. The bus ride.

Figure 7.2c. Pedestrianised shopping street.

Figure 7.2d 'Tour' end at Berri-UQAM metro station.

Figure 7.2'The trip'. All photographs by the author.

Figure 7.3a Montreal – downtown office buildings.

Figure 7.3b Montreal – downtown heritage buildings.

Figure 7.3c Montreal – downtown multi-storey apartment building.

Figure 7.3 Montreal's downtown environment. All photographs by the author.

The downtown environment of Montreal experienced included office buildings (Figure 7.3a) and shops, heritage buildings (Figure 7.3b) and residential complexes, like multi-storey apartments (Figure 7.3c).

Looking at the Social and Cultural elements in Montreal, the 'tour' provided a snapshot of the different strata in this slice of the city. As we moved through the city pedestrians ranged from shoppers in the more affluent west end of the city (Figure 7.4a and b) to a mixture of office workers and the homeless in a plaza nearby a metro station (Figure 7.4c).

Figure 7.4a Montreal – shopping precinct on Rue Sainte-Catherine.

Figure 7.4b Montreal –public transport users on Rue Sainte-Catherine.

Figure 7.4c Montreal –Montreal – plaza used by many diverse residents: from office workers to the homeless.

Figure 7.4 Montreal residents, viewed from the bus window. All photographs by the author.

Figure 7.5a Montreal – Commercial sector

Figure 7.5b Montreal –Upper-end shops on Rue Sainte-Catherine

Figure 7.5c Montreal – homeless outside Berri-UQAM metro station

Figure 7.5 Montreal street scenes. All photographs by the author.

The economic profile of Montreal could be seen through city buildings and citizens in the street. The trip began in Montreal's commercial centre, where many banks are headquartered (Figure 7.5a), then moved through upper-end shops on Rue Sainte-Catherine (one of the city's main shopping precincts) (Figure 7.5b), to the area adjacent to the UQAM metro station (Figure 7.5c), where many homeless people congregated.

Good governance can be seen through the activities and services provided by local and State governmental services. During the bus ride through part of Montreal

Figure 7.6a Montreal – Roadworks in Commercial sector .

Figure 7.6b Montreal – Road works on Rue Sainte-Catherine

Figure 7.6c Montreal – Government offices/services

Figure 7.6 Montreal government services. All photographs by the author.

the opportunity was taken to see what physical governmental activities and services could be seen. Immediately obvious were government maintenance services (Figure 7.6a and b) and government offices and services (Figure 7.6c). It was obvious that the local government was conducting many works programs and that services and information are accessible via prominent government offices. Here, the existence of multiple government buildings and numerous municipal works would seem to indicate good governance. However, the information gleaned is just surface information, and the effectiveness of provided governmental services and works were not gauged.

This quick preview illustrates the diverse nature of Montreal's central area.

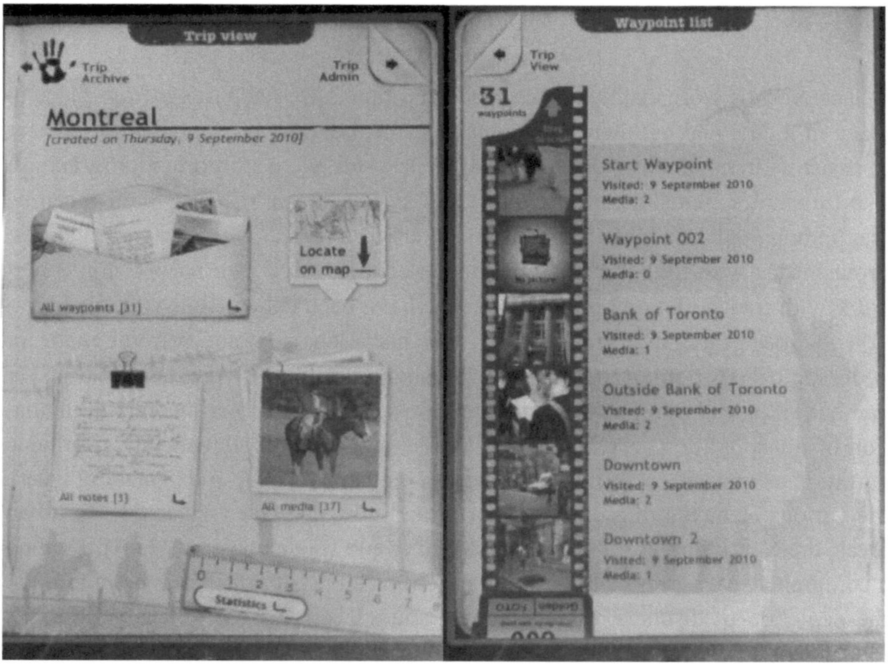

Figure 7.7 Trip Journal application

During the trip on the No. 15 bus service image collection was made using an Apple *i-phone 3* and the *Trip Journal* application (from iQapps). This application allowed for waypoints to be geolocated and each image georeferenced. Whilst the photographs alone provide rich information regarding the 'Triple Bottom Line +1", the additional geoinformation will allow for further assessment to be done. The *Trip Journal* application interface is shown in Figure 7.7.

Information graphics

The use of information graphics simplifies information and makes information access and consumption more efficient. Tufte (1990) has called graphics as "intelligence made visible". And, the increased use of information graphics is claimed to enhance how we better understand information, as supported by the statement below:

> *"The extent to which symbols and graphics affect our lives can be seen by the dramatic increase in IQ scores in all cultures which have acquired information technology: in the United States there has been an average increase of 3 IQ points per decade over the last 60 years, for a total of an 18 IQ point increase. There is no known biological explanation for this increase and the most likely cause is*

widespread exposure to text, symbols, and graphics that accompany modern life" (UW Computing & Communications, 2000, p. 2).

The world's workaday reality is space and time. We need to get from place to place in a city, and get there at certain times of the day. We invented timepieces to assist us in conducting our everyday lives. And, we use information graphics, like timetables, schedules, route maps and system maps to assist us in locating, wayfinding and understanding an urban environment. These representations, provision us with tools for planning and executing travel, so that we might best work with transportation conveyances. The have been called 'four-variable narrations of space-time' byTufte (1990). These narrations combine both the map, for geographical positioning and a notation of landmarks, like railway stations, tram car and bus stops and interchanges, and the time series, as a formal numerical tabulation of transportation services, in one usable, and readable, document. A graphic narration of transport systems can provide a valuable tool for understanding where the various elements of a city are located and how they are connected. In urban areas, the MetroMap_provides a record of time and spatial experiences – providing a comprehensive narrative description of a transport system. This is done through the depiction of distance, direction, time and speed. Basically, it simplifies the representation of systems, providing a tool for understanding.

Metromap metaphor

According to Donald A. Norman (1990), metaphors assist when aiming to design consistent objects and environments. The metaphor must be consistent for the task, tools and activities being undertaken. Spatial metaphors are used to represent and explain complex data sets. They allow the transformation of information into spatial representations affords easy reading and interpretation. The MetroMap metaphor has been utilised effectively to portray not just metropolitan transport networks, but also things like learning resources (Bang et al, no date), project plans (Scott et al., 2005) and for guided tours on the Web (Stott et al., 2005).

Perhaps the best example of how a MetroMap can simplify information is the Beck London Underground map of 1933 (Figure 7.8b). Beck's map simplified the portrayal of the complex transport system, making it more understandable than earlier representations, like the 1874 map, shown in Figure 7.8a.

Applying the metro map metaphor simplifies geographies, and makes them easier to interpret and understand. The MetroMap metaphor can be used to map 'other' geographies as well, like the map in Figure 7.9, which shows the relationships of different individuals and bands in the history of rock and roll.

The metro map that was used as a starting point for developing the annotated map for Montreal was the Societe de transport de Montreal (STM) map: *Plan du*

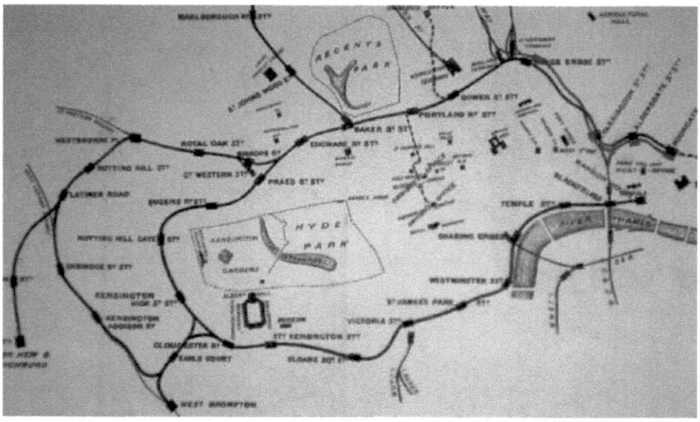

Figure 7.8a Early representation of the London Underground (1874). Source: http://blog.visualmotive.com/2009/automatic-generation-of-transit-maps/jpg.

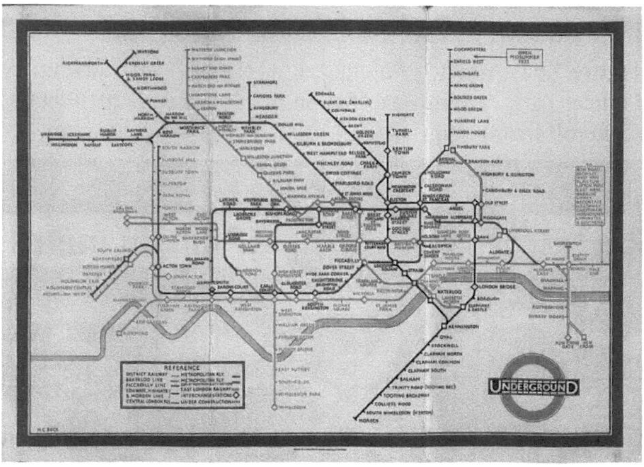

Figure 7.8b Harry Beck's London Underground map (1933). Source: http://uk.geocities.com/lhsoicher/images/1933a.jpg.

métro, shown in Figure 7.10. The map simplifies the geography of Montreal, so as to make the transport map 'work'. This map, and the stylized locations of the transport system and stations would form the basis to develop a map that could be annotated with Internet feeds. These feeds were structured to provide information that related to the four elements of the Triple bottom line +1.

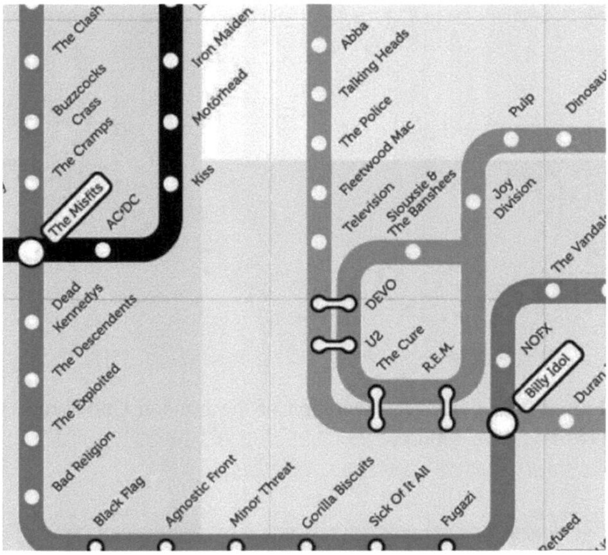

Figure 7.9a Alberto Antoniazzi - History of Rock‘n’roll Source: http://spacefiction.wordpress.com/2010/09/05/rocknroll-metro-map/

Figure 7.9b Montreal *Plan du metro* (2004)

Map Annotations

Maps delivered as printed paper products or computer screen images depicts snippets of information and facts by placing graphical beacons onto a myriad of line, point and areal icons, which graphically define the spatial 'place'. However, leveraging on contemporary communications systems, especially the Internet and Web 2.0 social software, elements from the 'real world' and human dialogue, including graphic dialogue, can be used to 'annotate' maps to give a real 'sense of place'. Input of additional information to annotate the map can add elements that are derived from emotional and perceptual viewpoints of real world information and defined by an individual or societal sense of place. In other words, an annotated map can let users consider human social and environmental complementary information, which are delivered through the process of (digital) map annotation to allow information to be seen from different and local viewpoints.

T.S. Eliot (1888-1965), in the poem *The Hollow Men* (1925) he wrote about the nature of existence (Brown, 2011). Part of the poem is included below:

"Between the idea

And the reality

Between the Motion

And the act

Falls the shadow"

We sometimes forget that we need to be reminded about how our ideas or aspirations actually occur in reality. And that our every initiative requires something to be implemented to achieve a desired result or to accord to a certain doctrine or vision. How do we determine if our desired result or vision has been affected and, what signals or signs should we seek for confirmation of success or lack thereof?

It is perhaps in the area of Eliot's 'shadow', somewhere between what a 'non-local' information gather perceives that a certain geography might actually look like - the idea - and the experienced reality of being or living in a place, that the user needs to be provided information about if a true understanding of a place is to be had.

When using maps, paper, digital or integrated media - if much more needs to be found about, then information like text and photographs can be consulted. Similarly, map collections, archival and current photographs, reports, files etc. can be investigated to compile a more complete 'picture' of the area being studied than just using maps alone.

Traditionally, we have used paper or electronic atlases to gain insight about an area. But, if a more comprehensive information resource is to be provided, then much richer and comprehensive methods for assembling an appropriate collection of pertinent information is needed. When developing integrated media applications, paper maps are replaced by computer graphics or pre-prepared maps (or screen combinations of both), text can be had as screen outputs or audio clips, pictures (still photographs, motion pictures (the 4th dimension), air photographs, satellite imagery and archival and current 'albums' via resources like *Flickr®*), and discrete or on-line data sets and Web feeds can all be integrated, composed, re-composed, displayed in 'user controlled' sequences or as 'stills on demand'. In other words an almost limitless 'decision-making' data base / information resource from which any number of required 'screen pages', rather than paper atlas pages, can be made available to specific users where they need that information to be made available. By 'annotating' the map with this additional rich information, and if this additional information could be presented about the environment of the city in the context of the Triple bottom line +1, perhaps a better (urban environmental0 'picture' of the city could be provided.

7.5 Building the Demonstration Prototype – The Annotated Map

Once we relied on written publications to furnish the information required to understand a city. These publications moved from paper to CD-ROM to the Internet and the Web. The range of titles now available has outstripped the 'old' printed collections of knowledge. Now, integrated media can be 'tapped' to provide access other information which can be used to verify or support information within a mapping application. Linking to commercially or publicly produced data repositories, locally or via some network, can provide a much more "media rich" and current product. Collections of 'facts' and current images are available via formal on-line services like newspapers, weather reports and scientific data collection Web sites. As well, informal sites, like Web Cams, *Twitter* ® and personal Internet-delivered information can complement the more 'official' information provided by formal environmental monitoring sites via more personal views of the city, neighbourhoods and 'personal spaces' in a city.

Visualising ideas

The idea for the demonstration prototype was to use the MetroMap as the key geospatial representation, about which appropriate environmental applications could be 'hung'. These annotations would provide rich information from feeds from newspapers, weather reports, *Twitter* ® feeds, etc. The environmental feeds would provide the real measure, or 'groundtruthing' (Cartwright *et al.*, 2001) about whether an idea or a vision for a society had been realized.

Groundtruthing information would be sourced from the elements of the 'Triple Bottom Line +1', described earlier. The concept was to have a very simple MetroMap, drawn with close reference to the Montreal STM metro map, which would be used for the 'first cut' product. The sketch used to develop this simple concept is shown in Figure 7.10.

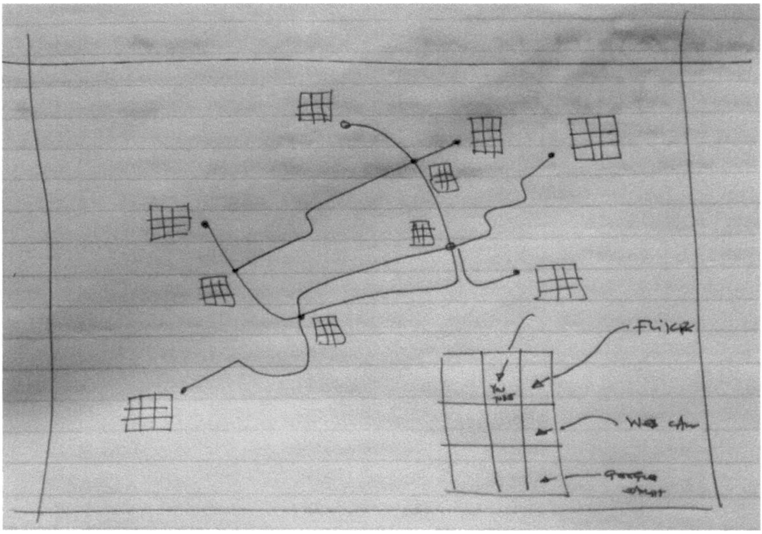

Figure 7.10 The City as a MetroMap concept.

To build the Web application, the STM metro map was re-drawn and simplified. This map is shown in Figure 7.11.

One criticism that can be rightly levelled here is that the actual metro map of Montreal is being used, where a more unique MetroMap should be used. However, as this was a demonstration prototype, the use of the re-drawn map provided a quick reference map. It is envisaged that when the demonstration prototype is further developed a MetroMap containing a level for each environmental indicator would be used, like the example for 'environment' shown in Figure 7.12.

Each of the four layers could be amalgamated to provide a comprehensive 'Triple Bottom Line +1' overview of the city (Figure 7.13).

Figure 13. The demonstration prototype was programmed in JavaScript. The brief was to provide four feeds per metro station node. When the user rolled-over the station symbol the feeds appear. These feeds are site specific and are self-generated according to the geo-location of both the metro line and the actual station. The full script for the package is provided in the appendix.

A 'localised' 'picture' of Montreal's environment is provided through 'self-generated maps'. These are products that are produced, after the initial design and

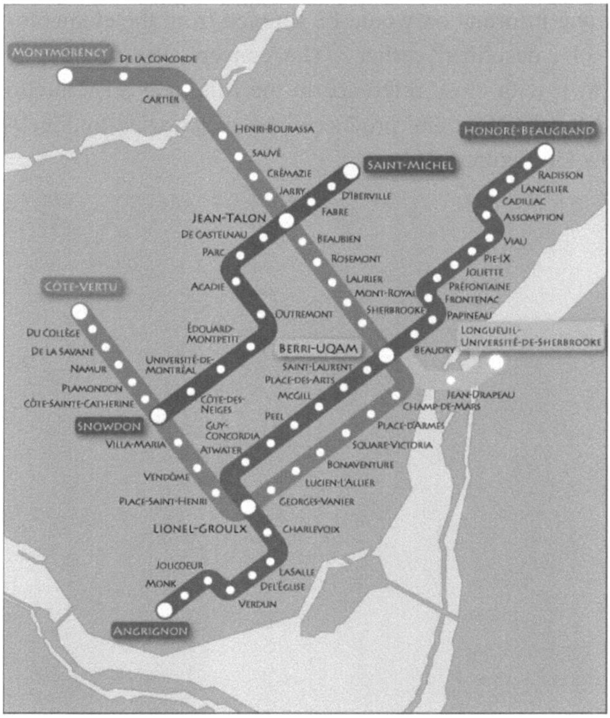

Figure 7.11 Montreal MetroMap base map.

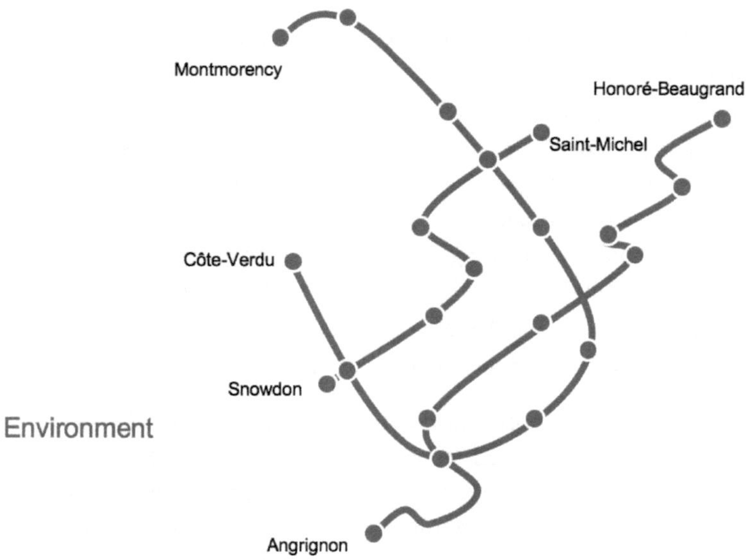

Figure 7.12 Montreal MetroMap metaphor and environment nodes (using metro stations).

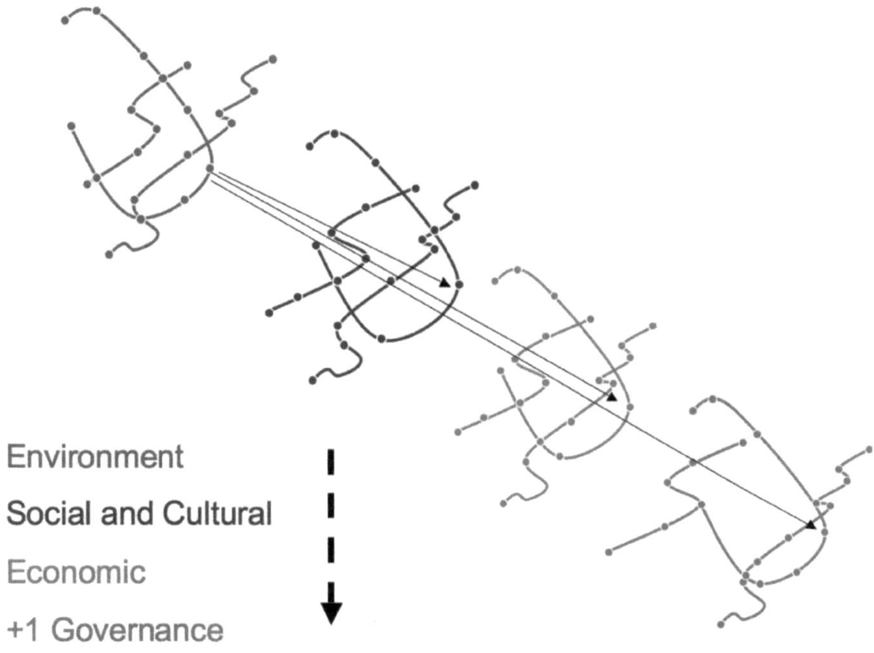

Figure 7.13 Montreal MetroMap metaphor and 4x layers of information.

production, with no, or little cartographer input. The maps can be generated from data that is sourced from large data repositories or collected in real time. A screen grab for the real-time feeds provided for the rollover for Honoré Beaugrand metro station, on the green line can be seen in Figure 7.14. Feeds are provided from the weather web site, *Twitter* ® feeds, the local newspaper and from *Flickr* (images). It is appreciated that there are differences in the accuracy of 'location' provided by social media like *Twitter* ®, *Flickr*® and *Facebook*®. They are essentially heterogeneous, which makes the assimilation of information from these various social media resources admittedly limited and relatively accurate. *Twitter* ® provides location via latitude and longitude and the information is limited to 140 characters. *Flickr*® images can be located via maps that are generated from Exchangeable image file format's (EXIF) standard tags for geo-location data. Cameras with on-board GPS can store locational information in the EXIF header. Here, locational information is as accurate as the on-board GPS. *Facebook*®'s focus is on 'Places and Deals', whereby it aims to link users of *Facebook*® to local facilities and services (David, 2011). Geolocation and *Facebook*® is generally limited to positioning *Facebook*® members using maps from *Google*® or *Bing*®. Whilst acknowledging that these information resources are heterogeneous, they nevertheless provide valuable, located information that would otherwise have no link to the geography of a city.

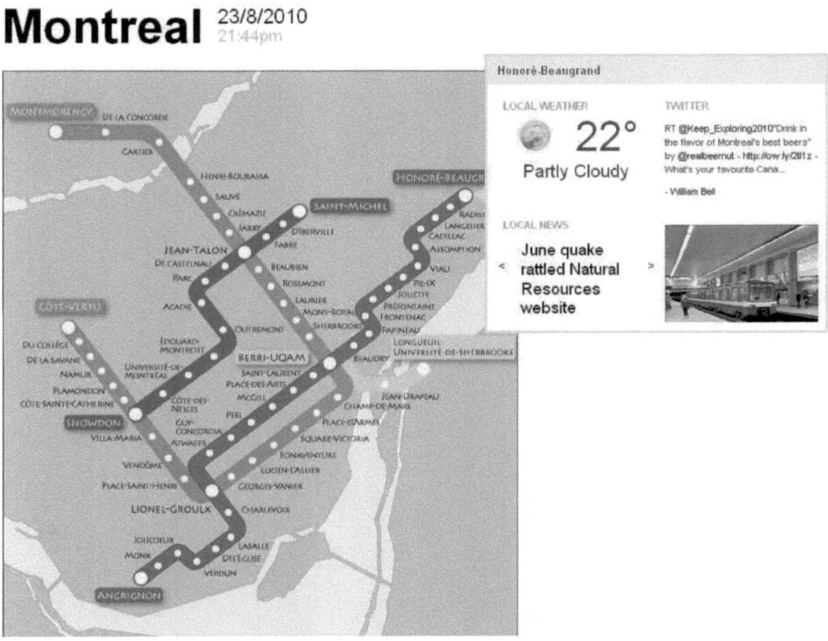

Figure 7.14 Annotated Montreal MetroMap and real-time information (here for Honoré-Beaugrand metro station area).

As well, there may be bias in the information provided. As well, some social software sites may attract users who may wish to conduct 'social mischief' by uploading erroneous or misleading information. For example, the #uksnow Map application (http://uksnowmap.com/) searches *Twitter* ® for real-time snow reports and maps composite locations. Users tweet their message with the hashtag #uksnow and add their location, which can be a postcode, suburb or town name (Field *et al.*, 2011). The application relies on the data providers to provide truthful information. However, many wrong reports were issued, undermining the site's credibility.

All feeds in the demonstration prototype are geo-location specific, here linked to the geo-location Honoré Beaugrand. Figure 7.15 shows the screen page generated when the mouse is rolled-over Còte-Vertu metro station.

Considering the Triple bottom line +1 elements the type of information feeds that could be incorporated are: environment – sensor feeds for air and noise pollution, social & cultural - *Twitter* ®, *Flickr*® and *Facebook*® pages showing social and cultural activities, economic – sites where government funds were being applied (roadworks, social housing, libraries, etc), and governance – feeds of (location-specific) articles from local papers commenting on the activities of local government.

Each information 'thumbnail' can be clicked and the information enlarged. This enhancement of the real-time information page is shown in Figure 7.16.

Figure 7.15 Annotated Montreal MetroMap and real-time information (Còte-Vertu metro station area).

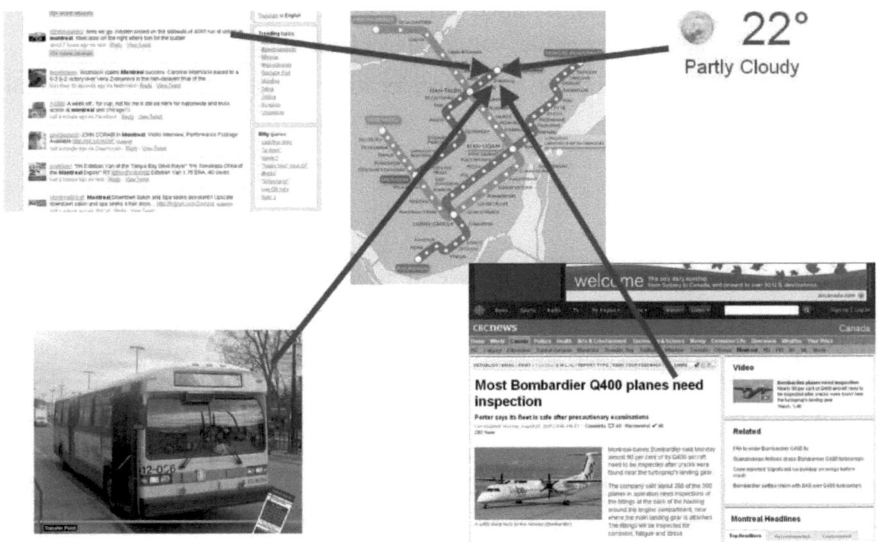

Figure 7.16 Annotated Montreal MetroMap information links.

The demonstration prototype can be viewed at the Web site at: http://prosimian.com.au/montreal/.

7.6 Further work – considerations

The work shown here is the 'first cut' of the demonstration prototype. It illustrates how real-time feeds can annotate maps, providing richer information. This is a work in progress, and the demonstration prototype will be tested, enhanced and improved. Present plans are to undertake further work around:

- Replacement of 'working' Montreal public transport metro map with project-specific '3+1' MetroMap. (This product would have metro 'lines' generated by linking points or areas of similar social or economic attributes. The lines would be in a state of constant flux, indicating how a dynamic city like Montreal changes through processes like gentrification of suburbs, urban renewal and the introduction of affirmative government actions.);
- 'Dragging' metro map nodes closer to relative zones of interest;
- Automatic generation of nodes with respect to clusters of indicators, rather than just stations (which will remain primary nodes);
- Generally testing the effectiveness of the metaphor model;
- Changes to scripting and sources of information
  - Identifying real feeds relative to triple bottom line +1;
  - Feeds more localised: weather, environmental monitors, community activities, city data (for example, valuations and building types); and
  - *Twitter*® feeds by content, rather than location (which tends to be too general)

7.7 Conclusion

The real problem about the effective use of an appropriate metaphor is to master the ways in which the metaphor should be best applied to the problem. A real understanding about how the metaphor should be applied is needed; so as to 'build' a metaphor that has relevance to the information being delivered and its timely delivery. When using geo-located annotated mapping products understanding the representational structure is central to the development of effective and efficient knowledge transfer information graphics. The information provided must he location-specific, location aware and rich. However, it is paramount that the product developed must be usable and attuned to the users' desire to be informed.

As Barlow (1995, p. 124) commented in the pioneer days of multimedia: "Marshal

McLuhan notwithstanding, it's (sic) the conversation, not the phone, that counts to my mother". The conversation here is the exchange of information from the 'real' members of Montreal's society to other members of their society, as well as its public servants, officials and leaders. Whilst the communication device here is the Web, and particularly the use of social software with Web 2.0 and using everyday consumer electronics. Barlow's comment still holds true to contemporary information communication: the tools must be appropriate for the user and the message must make sense, be appropriate and delivered in a timely manner for it to be useful.

Here, it is considered that a valuable tool can be provided to gauge the environmental 'health' of a city through the use of annotated maps that are built collaboratively using the many, varied resources that social software and Web 2.0 afford, and delivered through the use of effective information graphics, here the metro map metaphor. Information from everyday citizens, public repositories, various electronic media and feeds from individuals who go 'online' provide timely and important information. However, this power of being able to access these rich information resources is diminished if users are unable to consume and understand what can be somewhat complex and confusing information feeds. By better delivering geo-located information in a manner that affords improved comprehension the true value of this information can be exploited. Being able to capture, assemble and deliver information is not enough. It needs to be represented in a manner that is understandable and 'digestible. Here, the MetroMap metaphor was explored as a method whereby maps might be annotated with appropriate rich information, delivered effectively and made available to all.

Acknowledgements

Dean Sayers for drawing graphics of the Montreal Metromap.

Chris Marmo for completing the programming.

The three anonymous reviewers that provided valuable comments and suggestions for improving the manuscript.

References:

Bang, T., Grønbæk, K. and Hansen, P. St, no date, 'Using a Metro Map Metaphor for organizing Web-based learning resources'. http://www.gronbak.dk/kgronbak/publications/uploads/Bang-Gronbak-Hansen.pdf. Web page accessed 18 February 2011.

Barlow, J. P., 1995, "It's a Poor Workman Who Blames His Tools", Wired Scenarios, pp. 120 - 142.

Brown, P., 2011, T.S. Eliot's "The Hollow Men". Web page accessed 15 February 2011. http://www.suite101.com/content/ts-eliots-the-hollow-men-a340076

Cartwright, W., Crampton, J., Gartner, G., Miller, S., Mitchell, K., Siekierska, E. and Wood, J., 2001, "User Interface Issues for Spatial Information Visualization", CaGIS, vol. 28, no. 1, pp. 45 – 60.

David, M, 2011, "Where Location-Based Services are Taking Us: Brands & Beyond", siliconANGLE, November 4, 2011. http://siliconangle.com/blog/2010/11/04/where-location-based-services-are-taking-us-brands/ Elkington, J., 1997, Cannibals With Forks: The Triple Bottom Line of 21st Century Business.

Field, K, O'Brien, J. and Cartwright, W. E. 2011, "Exploring cartographic design in social-network map mashups", proceedings of the 24th International Cartographic Conference, Paris, France: International Cartographic Association, July, Heidelberg: Springer-Verlag.

Garland, K, 1994, Mr Beck's Underground Map, Capital Transport Publishing, Harrow Weald.

Gelernter, D., 1992, Mirror Worlds, New York: Oxford University Press.

Gelernter, D., 1994, "Highway of hope", The Age, April 19, p. 32.

GS@RMIT, 2010, Global Sustainability. http://www.rmit.edu.au/browse/Our%20Organisation%2FResearch%2FCapabilities%2FGlobal%20Sustainability%2FAbout%2FTriple%20Bottom%20Line%20Plus%20One/

Norman, Donald A., 1990, *The Design of Everyday Things*. New York: Doubleday.

Sandvad, K., Grønbæk, K., Sloth, L., and Knudsen, J. L., 2001, "A metro map metaphor for guided tours on the Web: the Webvise guided tour system", in *proceedings of WWW '01, the 10th international conference on World Wide Web.*

Stott, J.M., Rodgers, P., Burkhard, R.A., Meier, M. and Smis, M.T.J., 2005, 'Automatic layout of project plans using a metro map metaphor', in Information Visualisation, 2005. Proceedings. Ninth International Conference on Information Visualization, pp. 203 – 206.

Steinberg, S. G., 1997, "Mapping Science", Wired, January, p. 46.

The Economist, 2010, "A special report on smart systems", 20 pp.

Tufte, E. R., 1990, *Envisioning Information*, Cheshire, Connecticut: Graphics Press.

Appendix:

```
<!DOCTYPE HTML PUBLIC "-//W3C//DTD XHTML 1.0 Transitional//EN"
"http://www.w3.org/TR/xhtml1/DTD/xhtml1-transitional.dtd">

<html xmlns="http://www.w3.org/1999/xhtml"><head>

<!-- saved from url=(0014)about:internet -->

<title>Montreal Metro Map</title>

<meta http-equiv="content-type" content="text/html; charset=UTF-8">

<title>Uniting Care Community Options</title>

<link rel="stylesheet" href="Montreal%20Metro%20Map_files/reset.css">

<link rel="stylesheet" href="Montreal%20Metro%20Map_files/text.css">

<link rel="stylesheet" href="Montreal%20Metro%20Map_files/960.css">

<link rel="stylesheet" href="Montreal%20Metro%20Map_files/montreal.
css">



<!--[if Gecko 3]>

<link rel="stylesheet" href="<?=web_root?>/css/ff3.css" />

<![endif]-->

<!--[if IE 7]>

<link rel="stylesheet" href="<?=web_root?>/css/ie7.css" />

<![endif]-->



<script src="Montreal%20Metro%20Map_files/jquery-1.js" type="text/
javascript"></script>

<script src="Montreal%20Metro%20Map_files/jquery.js" type="text/
javascript"></script>

<script src="Montreal%20Metro%20Map_files/montreal.js" type="text/
javascript"></script>
```

```
<script type="text/javascript">

        $(document).ready(function(){

                        montreal();

                });

</script>

</head><body>


<div id="container" class="container_12">

        <div id="header">

                <h1>Explore Montreal</h1>

        </div>

        <div id="wrapper" class="grid_12">

                <div id="map">

                <img name="montreal_metro_map"

src="Montreal%20Metro%20Map_files/montreal_metro_map.gif"

id="montreal_metro_map" usemap="#m_montreal_metro_map" alt=""
border="0"

 height="700" width="600"><map name="m_montreal_metro_map"

id="m_montreal_metro_map">

                        <area shape="rect" coords="529,139,550,154"
href="javascript:;"

class="station " title="Honoré-Beaugrand" alt="honoreneaugrand">

                        <area shape="rect" coords="513,152,528,165"
href="javascript:;"

class="station " title="Radisson" alt="radisson">
```

```
                    <area shape=”rect” coords=”496,163,511,178”
href=”javascript:;”

class=”station ‟ title=”Langelier” alt=”langelier”>

                    <area shape=‟rect‟ coords=‟476,178,496,193‟
href=‟javascript:;‟

class=‟station „ title=‟Cadillac‟ alt=‟cadillac‟>

                    <area shape=‟rect‟ coords=‟472,198,488,214‟
href=‟javascript:;‟

class=‟station „ title=‟Assumption‟ alt=‟assumption‟>

                    <area shape=‟rect‟ coords=‟478,224,488,236‟
href=‟javascript:;‟

class=‟station „ title=‟Viau‟ alt=‟viau‟>

                    <area shape=‟rect‟ coords=‟458,235,474,250‟
href=‟javascript:;‟

class=‟station „ title=‟Pie-ix‟ alt=‟pieix‟>

                    <area shape=‟rect‟ coords=‟441,247,456,264‟
href=‟javascript:;‟

class=‟station „ title=‟Joliette‟ alt=‟joliette‟>

                    <area shape=‟rect‟ coords=‟423,265,437,274‟
href=‟javascript:;‟

class=‟station „ title=‟Préfontaine‟ alt=‟prefontaine‟>

                    <area shape=‟rect‟ coords=‟417,282,427,296‟
href=‟javascript:;‟

class=‟station „ title=‟Frontenac‟ alt=‟frontenac‟>

                    <area shape=‟rect‟ coords=‟416,301,436,318‟
href=‟javascript:;‟

class=‟station „ title=‟Papineau‟ alt=‟papineau‟>

                    <area shape=‟rect‟ coords=‟398,320,416,328‟
href=‟javascript:;‟

class=‟station „ title=‟Beaudry‟ alt=‟beaudry‟>

                    <area shape=‟rect‟ coords=‟479,343,499,364‟
```

```
href="javascript:;"

class="station „ title="Longueuil-Université-de-Sherbrooke"

alt="longueuil">

                              <area shape="rect" coords="436,364,453,377"
href="javascript:;"

class="station „ title="Jean-Drapeau" alt="jeandrapeau">

                              <area shape="rect" coords="370,335,392,354"
href="javascript:;"

class="station „ title="Berri-Uqam" alt="berriuqam">

                              <area shape="rect" coords="382,379,398,395"
href="javascript:;"

class="station „ title="Champ-de-Mars" alt="champdemars">

                              <area shape="rect" coords="356,395,375,413"
href="javascript:;"

class="station „ title="Place-D'Armes" alt="placedarmes">

                              <area shape="rect" coords="334,414,347,430"
href="javascript:;"

class="station „ title="Square-Victoria" alt="squarevictoria">

                              <area shape="rect" coords="305,430,324,449"
href="javascript:;"

class="station „ title="Bonaventure" alt="bonaventure">

                              <area shape="rect" coords="281,453,298,470"
href="javascript:;"

class="station „ title="Lucien-L'Allier" alt="lucienlallier">

                              <area shape="rect" coords="348,303,363,320"
href="javascript:;"

class="station „ title="Sherbrooke" alt="sherbrooke">

                              <area shape="rect" coords="334,287,347,302"
href="javascript:;"

class="station „ title="Mont-Royal" alt="montroyal">
```

```
                       <area shape="rect" coords="322,265,334,281"
href="javascript:;"

class="station „ title="Laurier" alt="laurier">

                       <area shape="rect" coords="303,246,322,264"
href="javascript:;"

class="station „ title="Rosemont" alt="rosemont">

                       <area shape="rect" coords="289,225,305,240"
href="javascript:;"

class="station „ title="Beaubien" alt="beaubien">

                       <area shape="rect" coords="356,350,368,368"
href="javascript:;"

class="station „ title="Saint-Laurent" alt="saintlaurent">

                       <area shape="rect" coords="326,368,347,382"
href="javascript:;"

class="station „ title="Place-Des-Arts" alt="placedesarts">

                       <area shape="rect" coords="305,388,324,405"
href="javascript:;"

class="station „ title="McGill" alt="mcgill">

                       <area shape="rect" coords="281,405,298,423"
href="javascript:;"

class="station „ title="Peel" alt="peel">

                       <area shape="rect" coords="258,423,273,438"
href="javascript:;"

class="station „ title="Guy-Concordia" alt="guyconcordia">

                       <area shape="rect" coords="234,438,248,456"
href="javascript:;"

class="station „ title="Atwater" alt="atwater">

                       <area shape="rect" coords="259,473,273,488"
href="javascript:;"

class="station „ title="Georges-Vanier" alt="georgesvanier">

                       <area shape="rect" coords="234,482,253,507"
```

```
href="javascript:;"

class="station „ title="Lionel-Groulx" alt="lionelgroulx">

                        <area shape="rect" coords="253,512,272,528"
href="javascript:;"

class="station „ title="Charlevoix" alt="charlevoix">

                        <area shape="rect" coords="257,539,272,554"
href="javascript:;"

class="station „ title="Lasalle" alt="lasalle">

                        <area shape="rect" coords="240,554,255,571"
href="javascript:;"

class="station „ title="De l'Église" alt="deleglise">

                        <area shape="rect" coords="218,568,234,586"
href="javascript:;"

class="station „ title="Verdun" alt="verdun">

                        <area shape="rect" coords="196,562,211,575"
href="javascript:;"

class="station „ title="Jolicoeur" alt="jolicoeur">

                        <area shape="rect" coords="173,576,189,588"
href="javascript:;"

class="station „ title="Monk" alt="monk">

                        <area shape="rect" coords="147,586,171,608"
href="javascript:;"

class="station „ title="Angrignon" alt="angrignon">

                        <area shape="rect" coords="204,473,218,488"
href="javascript:;"

class="station „ title="Place-Saint-Henri" alt="placesainthenri">

                        <area shape="rect" coords="183,449,204,463"
href="javascript:;"

class="station „ title="Vendôme" alt="vendome">

                        <area shape="rect" coords="165,423,181,438"
href="javascript:;"
```

```
class="station „ title="Villa-Maria" alt="villamaria">

                            <area shape="rect" coords="145,396,165,415"
href="javascript:;"

class="station „ title="Snowdon" alt="snowdon">

                            <area shape="rect" coords="135,380,149,396"
href="javascript:;"

class="station „ title="Côte-Sainte-Catherine"
alt="cotesaintecatherine">

                            <area shape="rect" coords="118,366,135,380"
href="javascript:;"

class="station „ title="Plamondon" alt="plamondon">

                            <area shape="rect" coords="106,346,125,362"
href="javascript:;"

class="station „ title="Namur" alt="namur">

                            <area shape="rect" coords="94,328,109,343"
href="javascript:;"

class="station „ title="De La Savane" alt="delasavane">

                            <area shape="rect" coords="82,312,94,328"
href="javascript:;"

class="station „ title="Du Collège" alt="ducollege">

                            <area shape="rect" coords="65,293,89,310"
href="javascript:;"

class="station „ title="Côte-Vertu" alt="cotevertu">

                            <area shape="rect" coords="177,371,196,388"
href="javascript:;"

class="station „ title="Côte-des-Neiges" alt="cotedesneiges">

                            <area shape="rect" coords="210,350,224,366"
href="javascript:;"

class="station „ title="Université de Montréal"

alt="universitedemontreal">

                            <area shape="rect" coords="235,328,255,343"
```

```
href="javascript:;"

class="station „ title="Édouard Montpetit" alt="edouardmontpetit">

                        <area shape="rect" coords="248,295,266,312"
href="javascript:;"

class="station „ title="Outremont" alt="outremont">

                        <area shape="rect" coords="224,265,240,281"
href="javascript:;"

class="station „ title="Acadie" alt="acadie">

                        <area shape="rect" coords="227,240,243,255"
href="javascript:;"

class="station „ title="Parc" alt="parc">

                        <area shape="rect" coords="251,225,266,237"
href="javascript:;"

class="station „ title="De Castelnau" alt="decastelnau">

                        <area shape="rect" coords="270,203,292,225"
href="javascript:;"

class="station „ title="Jean-Talon" alt="jeantalon">

                        <area shape="rect" coords="298,188,313,203"
href="javascript:;"

class="station „ title="Fabre" alt="fabre">

                        <area shape="rect" coords="315,174,334,189"
href="javascript:;"

class="station „ title="D'Iberville" alt="diberville">

                        <area shape="rect" coords="334,151,356,176"
href="javascript:;"

class="station „ title="Saint-Michel" alt="saintmichel">

                        <area shape="rect" coords="255,183,270,198"
href="javascript:;"

class="station „ title="Jarry" alt="jarry">

                        <area shape="rect" coords="242,163,255,176"
href="javascript:;"
```

```
class="station „ title="Crémazie" alt="cremazie">

                          <area shape="rect" coords="226,143,242,159"
href="javascript:;"

class="station „ title="Sauvé" alt="sauve">

                          <area shape="rect" coords="210,121,226,137"
href="javascript:;"

class="station „ title="Henri Bourassa" alt="henribourassa">

                          <area shape="rect" coords="169,71,192,93"
href="javascript:;"

class="station „ title="Cartier" alt="cartier">

                          <area shape="rect" coords="109,61,128,82"
href="javascript:;"

class="station „ title="De La Concorde" alt="delaconcorde">

                          <area shape="rect" coords="51,61,73,82"
href="javascript:;"

class="station „ title="" alt="montmorency">

               </map>

               </div> <!-- map -->

        </div> <!-- wrapper -->

</div><!-- container -->

<div qtip="67" class="qtip qtip-light" style="-moz-border-radius: 0pt

0pt 0pt 0pt; position: absolute; width: 400px; padding: 0px 0px 0px

14px; display: none; left: 329px; top: 10.5px; z-index: 6000;"><div

class="qtip-tip" dir="ltr" rel="leftMiddle" style="position:
absolute;

height: 14px; width: 14px; margin: -7px auto 0pt; line-height: 0.1px;

font-size: 1px; left: 0px; top: 50%;"><canvas height="14"
width="14"></canvas></div>

  <div class="qtip-wrapper" style="position: relative; overflow:
```

```
hidden;

 text-align: left;">    <div class="qtip-contentWrapper"

style="overflow: hidden; background: none repeat scroll 0% 0% white;

border: 3px solid rgb(226, 226, 226);"><div style="background: none

repeat scroll 0% 0% rgb(241, 241, 241); font-weight: bold; padding:
7px

12px; color: rgb(69, 69, 69);" class="qtip-title">Montmorency</div>

  <div style="background: none repeat scroll 0% 0% white; color:
rgb(69,

 69, 69); overflow: hidden; text-align: left; padding: 5px 9px;
height:

250px;" class="qtip-content qtip-content"><img

src="Montreal%20Metro%20Map_files/metro.jpg"></div></div></div></
div></body></html>
```

Section II

Mapping From the Ground: The Artistic/Humanist Perspective

Chapter 8

Roaming Montréal: Seeking the Representation of the 'Geographic Self'

Laurene Vaughan

Associate Professor in the School of Media and Communication and Research Leader within the Design Research Institute at RMIT University, Melbourne

8.1 Introduction

My first encounter with the data and the database that was at the centre of the workshop 'Mapping the Environmental Issues of the City,' was at first disheartening and frustrating. I am not familiar with the workings and creation of databases. The data of my world is not ordered in the way that data emerges from a database. I work in the space between interpretation and proposition, and the data of my practice emerges from a variety of messy sources, usually in the form of texts, people, conversations and observations. The process of accessing the data is one of discovery, laying it out and peering in. The data rarely looks the same twice; it is always in a state of flux being closely aligned to the actions of the everyday. When the environmental data of Montréal arrived I couldn't find my way into the place of such statistics. The digital files wouldn't open, I didn't know the file types; it was a completely foreign encounter for me. The language, the form and the location, I tried to find the synergies. I tried to see the city, but all I could see were statistics, colour gradients creating outlines of regions, each a reference to unknown locations and inhabitants. I felt locked out, once removed not just from the place, but also from this way of knowing and representing that place.

Boundaries are an essential part of place-making and place-finding, they are a central feature of geography; lines are always being drawn to mark the division between here and there, or this place and that. In this case it was the technology of the database, which created and highlighted for me that a barrier existed between me, and the place I was seeking to know. Finding my way through or around this block became my central interest in my project response, *Roaming Montréal.* As I ventured into the project, I came to understand that this initial database block, was just one of the many barriers that stood between this city and me. When I started the project I was extremely ignorant about Montréal. I knew it was on the eastern side of Canada; I knew it was part of Quebec; and I knew that the people spoke

French. It was not until I bought maps and guidebooks, that I started to gain some level of understanding of the city, and still it was only a touch on the surface. The various data sources that I had collected; the maps, guidebooks and environmental databases were abstract bridges to the city for me. Through them I began to have an idea, but I did not have any real knowing of this place.

The objective of the *Roaming Montréal* project was to explore the relationship between the abstract 'data' of a city and its relationship to the actual practices of inhabitation by the residents of that city. I aimed to focus on the tensions between the objective and the subjective, the collective and the individual, and of course the public and the private. Consequently I wondered: *how do we live the contents of a database? And, how can we 'see' the database of the city in the everyday places and actions of the people of the city?*

8.2 The Problem with Data and Representation

Whenever we attempt to render and record a city we are always challenged by the differences between what is shown and the lived reality of a place. For no matter how accurate the facts that inform the representation are, they are by their very nature abstract, by virtue of being once removed from the reality of the location. The data that we collect and hold within various databases are small isolated 'bits' that can be ordered and reordered according to the system (the database) that we put them within. They are at least once removed from their source of origin, and this has implications for what that data can tell us about our everyday understanding and experience of the world; particularly when that world, is not our own.

Typically we understand data as being a set of facts or statistics about something or somewhere; these data are the measurable, collectable, countable and storable information about the world. These facts pertain to both the natural, constructed and inhabited world, and they have an element of truth and certainty about them. The database is the location where these facts are housed; the facts sit there, waiting for our intervention in order to come to life for some use or another. Mapping and other forms of visualization are amongst the methods that are used to transform the data into something that can have meaning for specific use by specific people; it is hoped that this data will help us to understand or make sense of something. The relationship between the 'data' (the details) and the database (the system that manages the bits) exists outside the reality of the world that it represents, and at the same time it can be argued that these 'bits' of information *are* the truth of the world as known at the time of collection by those that collect.

Underpinning this project is a foundational belief that no matter how accurate, how aesthetic or how inclusive, any representation of place is by nature of being a representation, limited and quite possibly not of that place. As argued broadly in the literature place is an outcome, a manifestation of the practices of habitation, it is a lived entity, thus any or all representations are always interpretations. Representations are constructions; they are the re-presentation of something for others to engage with. Each component of the construction is made, manipulated, and organized to meet a certain need or expectation; thereby representing a diverse array of actions, intentions and selections that have been done in order to realize an outcome. As noted by Tufte (1997, p. 25), no graphic representation is ever socially, culturally or politically neutral, and new technologies make this even more so.

> *Enthusiasts, partisans, and liars have long tinkered with graphical evidence by dequantifying images, selecting and hyping advantageous visual effects, distorting data. Recently, inexpensive computing and ingenious techniques for image processing have provided endless new opportunities for mischief. Arbitrary, transient, one-sided, fractured, undocumented materials have become the great predicament of image making and processing. How are we to assess the integrity of visual evidence? What ethical standards are to be observed in the production of such images?*

It is not my intention to question the integrity of the data or the database that was designed by my colleagues to be the central focus of this workshop. I have no doubt that they worked with the utmost integrity, for it is not this specific data and database that I am questioning, rather it is the notion that any such information can provide us with any and definitely not absolute truth about a place. It is well known that it is impossible for any map to represent all the data that pertains to it, or for it to ever be complete. As Janet Abrahams and Peter Hall state in their introduction to *Elsewhere Mapping*, the term'(m)apping refers to a process - ongoing, incomplete and of indeterminate, mutable form' (2006, p.12). This process is one that is always seeking new ways to visualize and represent data, and often 'it is not so much what a map shows as what it omits' (p.13) that is central to the representation. Limited by the vagaries o time, mapping is a process that is always out of date, for there will inevitably be a gap between the data that is recorded and the speed within in which it can be mapped.

Our inclination can be to engage with a map, like the data that informed its creation, as a resolved rather than a dynamic entity. Its 'artefactualness' in form, its beauty and mystery can seduce us (even though we know it to be flawed) into a state of complacency where by we perceive it as being complete. Armin Lobeck (1993) in his text *Things Maps Don't Tell Us*, explores the limitations of representation arguing that there are two important components for engaging with a map. First there is the act of map reading, and secondly, there is map interpretation. He explains this as:

> *Map reading is what all of us do when we want to find out where a place is on the*

map, or the distance between places, or their relative positions or any other simple geographical fact. Map reading is what scouts do when they use maps out of doors to keep from getting lost. Map reading is what the motorist does when he uses a road map to find the best route for his journey.

But map interpretation is much more than all of this. Map interpretation is like the process of reading between the lines of a story where by the reader draws certain inferences and conclusion, which the author did specifically make. On a map for example, the form of the hills, the location and pattern of rivers, the outline of the coast, and other details may provide the clue to certain geological information which the maps makes no pretense of giving. So, in this book, we shall take certain simple geographical information, which we can read from the map and see what influences we can draw from it, to interpret and explain how these came about. (pp. xi – xii)

Lobeck's distinction between reading and interpreting provides a useful framework for my concerns about the limitations of data and databases for enabling us, no matter how we go about our configurations, to really know a place through such means. For example the data that was collated for the environmental database of Montréal focused on various issues and aspects that inform our experience of that city. Therefore in a case such as this, data that is related to health, may show us that there are certain incidents of certain kinds of health related issues in certain parts of the city; and from this we can know that in area X there is a high proportion of Y, and in contrast in area B, no amount of Y has been noted. Consequently we can say that element Y is present in area X but no sign is recorded in area B. This gives us one level of knowing about the element and these two specific sites within the city. We could then seek out other data about areas X and B and start to build a picture of the city through fractured parts which we hope will enable us to gain a richer picture of what is taking place; this is an act of interpretation. To move beyond surface interpretation we must, as Lobeck says, read between the lines and start to imagine, to hypothesize about the threads that bind these elements together (why X and not B); and in so doing we start to make a shift from being in a space of fact to one of fiction. We might say, that we start to get a glimpse of the lives that live the data in place.

The act of interpretation in the reading of data is in itself a practice of habitation and familiarization reminiscent of Michel de Certeau's (1984) *strategies* and *tactics* for the practice of everyday life, and of everyday place-making. Interpretation is a tactic for starting to know about the unknown and perhaps the unfamiliar place. For a tourist and outsider such as myself, it is one way that I can begin to encounter a foreign place, and start to create my own narrative of the city, flawed and ill informed as it may be. Let us return to place X and the data of proportion Y, from my reading of the data I only know that Y exists, the strategies and tactics that are employed by the people of place X to deal with factor Y are absent. I have no idea what the presence or absence of this element means in the everyday. These pieces of truth (data) that are used to represent or describe a particular entity within a location of the world, are only partial truths, outside of the context of meaning and localized action, they tell us nothing about the lived truth of the experience of the world of those that inhabit area X.

As I engaged with the databases central to this project, I as a non-cartographer was left outside of the data. This wasn't only because of my inability to understand the science of the terminology, nor was it because I am also not a user of the technology that ran the database; rather my real gap was in my lack of knowledge of this place. I could not reconcile the data, the large spread sheets that covered my studio floor, with the reference map that accompanied it, nor with the maps I had bought of the city. My mapping practice is not one of objective data representation, mine is a mapping practice of the phenomenon of place and the inhabitation of place. This gap between the data on the city of Montréal and myself is symbolic of what I see as the gap between the facts, their representation and the experiences of people.

8.3 Roaming Montréal

How do such facts really exist in the lived city?

Figure 8.1.

Following on from my initial data block, I decided that my objective in this project would be to seek out the living database of Montréal. As a phenomenologically orientated artist and designer, the combined abstraction and certainty of the data and the various configurations of the database maps, frustrated my desire to know some of what it is like to be in a place.

My initial response of collecting maps and guidebooks was my way of trying to make sense of the socio-cultural landscape of the Montréal. This was my view from a far. On arriving in Montréal I questioned a colleague about the city, the history and who lives where (see Figure 8.1). I knew from the literature that Montréal has a rich cultural history, and I wondered how does the Anglo and French divide realize itself in the city? What are the territories? What are the behaviors? And as I read and talked I discovered that this city is much more than its popular bi-partisan divide. For underneath these broad headings of French and English, there are many other groupings, traditions and homelands. Montréal is a colonized territory, and with this some cultures come, some go, some flourish, and some struggle or disappear.

Figure 8.2.

In order to seek out the living database of this rich cultural landscape I decided to employ the methodologies of walking and noticing as a means for starting to discover this place (Vaughan 2009). This is a methodology that I have used in a number of other projects, as it enables the discovery of localized knowledge about somewhere and it accepts the fundamental bias of the observer in forming a narrative about a place (Vaughan & Akama 2009, Vaughan 2009b). My practice of 'noticing' is derived in the work of John Mason (2002) who conceived of it as a discipline for individual discovery (see Figure 8.2). What one person notices will be unique to them, even though many may notice the same thing. *Roaming Montréal* adopted a directed route of noticing. Similar to the Dada and Situationist *dérives*, I set out on a walk through an identified area with the intention of seeing what I would see (see Figure 8.3). The selected region for this investigation was *Griffintown.* I selected this part of the city because although still named after its past inhabitants, this particular cultural group has become lost in the narrative of the city, as it has transformed over time.

Figure 8.3.

Griffintown is a socially contested area within the city of Montréal. Being close to the downtown area, it is like so many areas world over, in a state of transition from being an area of factories and workers houses, to becoming one of office blocks, condominiums and expressways. With this knowledge, a colleague I went for a walk in search of *Griffintown* in the present, whilst also wondering if we would see signs of the past, or any traces of the data that I had collected from the various sources.

The outcome of this walk is a visual essay that shows fragments of life in a city (http://roamingmontréal.posterous.com). The photographic essay features details of buildings, evidencing signs of time and wear; of signage and street names, some hark back to old times, others' name changes marking future plans. As I encountered this part of the city, I endeavored to walk and photograph in a way that was in keeping with the methodology of noticing and not looking. Noticing requires a light touch, a gaze or a glance, an awareness of what passes our peripheral view; this is a challenging way to view a place, for our inclination is to focus and the more we notice, the more we see.

In the midst of our meandering we happened upon small points of protest placed upon the city streets by local residents. These small sites were chanced upon, and were striking not because they made use of bold graphics or harsh slogans, rather these were protests on the personal scale, utilizing elements of the domestic world lifted out of the house and on to the street, highlighting that this was a dwelling place, where people lived and despite the ambitions of town planners and developers this was already someone's home. The makers of these artifacts were absent, there was a sense of walking into a place just as everyone left, or onto a stage waiting for a performance to being; and yet even in their absence the strength of their convictions hovered like ghosts leaning against walls. This highlighted for me the significance of the elements that mark the habitation of a place and it is these, whether we view them as a local or an outsider, that make the richness of that place. It was this lifeblood of desire and distress, that was the living data that I sought, and it emphasized for me, our inevitable inability to represent this depth of place in any form of image, number or text (see Figure 8.4).

Figure 8.4.

8.4 Finding and Representing Place

How can we read through data to see such complexity about the living of place? The phenomenon of place involves much more than a series of geospatial markers of a location set in space. As Relph states place is more then the 'where of something' (1976 p. 3). Place is a localized activity. Place is both geographically specific and it is also practiced through the actions of the everyday (de Certeau 1984). To conceive of the making, recording or representation of place(s) is to engage with the actions of people as they make their way through the socio-cultural and political landscapes that we have made. Place is a complex entity, comprised of multiple forms. Relph cautions against trying minimize or unify a definition, or perhaps even a description, of place. Arguing instead that we should embrace the complexity of place and seek clarity through that complexity. 'By taking place as a multifaceted phenomenon of experience and examining the various properties of place, such as location, landscape and personal involvement, some assessment can be made of the degree to which these are essential to our experience and sense of place' (1976, p. 29).

A phenomenological exploration of place such as *Roaming Montréal* calls us to consider the nature of the experience of space and what it is that makes place, place. For many geographers and philosophers within this tradition it is the concept and the acts of dwelling, which are essential to our making and comprehension of place. These acts of dwelling include methods of way-finding, way-showing and other representational artifacts which help to translate and enable our experience of place. Our knowledge of place is intricately local; it is localized to the experiences and perceptions of each individual, and their perceptive parameter. For the city dweller, the reality of the city is the street, for the rural dweller it may be the limits of the property or town. Local places are those that are within the realm of our understanding and experience of the everyday, and they may be dwellings both within and outside of the home. As Bachelard argues, 'our home is our corner of the world... it is out first universe, a real cosmos in every sense of the world' (1969, p. 20) and this home is the first parameter of the local, and our world expands from there.

One of the challenging aspects of place is that it is in a constant state of evolution. 'Whether place is understood and experienced as landscape in the direct and obvious sense that visual features provide tangible evidence of some concentration of human activities, or in a more subtle sense as reflecting human values and intentions, appearance is an important feature of all places' (Relph 1976, p.31). Place is not static, and therefore all or any representation of place will be limited, and the data at its source will be outdated. Although multiple people inhabit a place, the phenomenon of place is distinctly individual and communal or social. Thus another layer of complexity emerges in the discussion of the representation of place, for if we are to conceive of place as *practiced space* (de Certeau 1984) and that it is these practices and the social or communal context that they occur within in that create a place, then, how can static data, adequately record or convey such subtle complexities?

According to Relph, '(t)he identity of a place is as much a function of intersubjective intentions and experiences as of the appearances of buildings and scenery, and it refers not only to the distinctiveness of individual places but also to the sameness between different places' (1976, p. 44). These are the social and physical attributes of place, but what of the representation of place? Each of these entities is made up of smaller pieces, especially when we expand to thinking about an entity as complex as a city place. The city as place has many diverse and complex attributes. These include, the inhabitants, the social and cultural structures and the sub-places (what we might call a suburb or borough within a larger city place entity). Each of these has its own individual qualities, the respective individual inhabitants will know each in their respective ways, and yet there is also a layering of communal identities and interpretations that then bridge the individual self with the we. Relph calls this the 'sameness in difference' in our understanding of place (1976, p.45). This is a sameness, which is informed or framed by the nature of the connection between the self to the place; and/or the connection between the self, and the community of that place.

What are the boundaries around the status of inhabitant of a place? When am I a local, a visitor or a tourist? At what point does my status shift and then with that, how does this then shape and inform my sameness in difference to the community of a place? When I am endeavoring to read a place, my connection as either an insider resident or an outsider tourist will frame my reading of the entity and it's identity. 'To be inside a place is to belong to it and to identify with it, and the more profoundly insider you are the stronger is this identity with place' (Relph, 1976, p. 49).

The status of being either inside or outside of place is dynamic. It is always poised in a position of change, and with time and connection our status of insider can deepen. By being absent from a place we can transition from being deeply inside to being, or slowly moving towards the periphery of the external boundary line. As the focus of our intention shifts so does our placement in relation to the boundary.

In *Roaming Montréal* I embraced the position of outsider. I was a tourist, a transient member of a place only able to engage with a surface reading of what the city was and the practices of people living there (see Figure 8.5). Over the span of four days I attempted to gain some knowledge of the city, and with time the structures and practices of the city increased in familiarity to me. The topology and landmarks, cafes, the location of the university, the underground path of the subway all became familiar; these things started to transition into my localized daily activity, but I was not a local, I was to be in the city for less than one week. My position to the boundary was shifting, but it was still firmly on the side of the outsider.

8.5 The Geographic Self and the City

Figure 8.5.

The central premise of human geography is that we know natural and cultural landscapes through the dimension of human experience. The philosopher geographer Edward Casey extends this understanding through his argument that place is the outcome of the 'geographic self' (2001, p. 683). This is a term that he has developed in response to Sacks (1997) discussion of *Homo Geographicus*, and it enables Casey to explore 'the nature of the human subject who is orientated and situated in place' (ibid). This is an argument, which is itself situated within the phenomenological discourse of place. Sacks and Casey are both exploring the web of influences and interconnections between nature, culture and human beings; and in this space it is the human, the body or the self that is the essential element to the making of place.

> *The self has to do with the agency and identity of the geographical subject; body is wheat links this self to lived place in its sensible and perceptible features; and landscape is the presented layout of a set of places, not their mere accumulation but their sensuous self-presentation as a whole.* (Casey 2001, p. 683)

When the geographic self seeks to position itself within place, it does this not only by considering the physical attributes of the locale but also through a framework of relating; relating to others and of being related to (Relph 1976 p.39). It is in this way that place should be conceived of as an inter-relational entity involving exchanges between people, artifacts and locations all within the context of time. These are the construct of our intentions, our attitudes, purpose and experience, and underpinning all these are the practices of everyday life.

For Casey it is Pierre Bourdieu's notion of *habitus* that best articulates this link between the practices of life and our connection to space and the subsequent making of place. He states, 'I propose that habitus is a middle term between place and self - and, in particular, between lived place and the geographical self. This self is constituted by a core of habitudes that incorporate and continue, at both psychical and physical levels, what one has experienced in particular places' (Casey 2001, p. 686). As the middle space the practices of habitus are the things that enable place to be made in the external landscape and in our internal knowing and embodiment of place. These practices include the various practices from work to leisure, the spectacular and the mundane, and together they comprise the many ways that we know, relate to and make 'our' place(s) in the world. Place in this context is both geospatial (for we are always located somewhere in space), but it is also place understood as the connected space or location of life. This is place as meaning and meaning making within the locales of our existence. Inhabitation is 'placial' and temporal, and results in the place-world that is both perceived and conceived through our active engagement with the world.

The only way that we can know the lived world is through the body. It is through our bodies that we live and know and act out the world. For Casey this happens in two ways: outgoing and incoming. 1. Outgoing is when the lived body encounters the place-world by *going out to meet it*. In this way there is an interconnection between the body's actions in the world, which are framed by the form of that world and then created through the actions of the body in that world. 2. Incoming is when the body doesn't just go out to meet the world, but when the body then bears the marks of that world on itself. Who we are is shaped by the world that we live in.

It is in this way that *homo geographicus* or the *geographic self* is formed by the place-worlds it inhabits. No matter how long we are in place, if there is a connection to such a place we are changed, and we bear the marks of such places on our being. We are an expression of place, and as argued by Casey (2001, p. 688) *habitus* does not convey the depth of this connection to place that is realized in our being. For although we make the world, the place-world is both individual, my particular knowing of place, but also communal, we make the communal place world and this world is always in flux, and the traces of our individual or collective presence can be removed (albeit this may take time and re-appear). The geographical self is the self that has been and continues to be made through our encounters with the place-world.

> *Just as there is no emplaced self except as a body/self, there is not place either without such a self. There may well be space and location in the absence of an embodied self, but in the presence of place there can be no subject capable of possessing habitus, undertaking habitation, and expressing the idiolocality of place itself.*
>
> Casey p. 689

Casey's proposition that the making of place is an interdependent activity where by we both make place and make ourselves through our presence in place, adds a layer of ambiguity to our position as insider or outsider. For the relationship implies, that no matter how long we are in a place, it, and we are changed by our presence. I as a tourist walking the streets of Montréal, seeking out the city am not a passive observer. My presence in itself adds to what that city is, and I too am shaped in response to having been there. As I walk the city I am also molding myself in some new way.

8.6 The City, Place and Me

Cities are located within ecological systems, even when such 'natural' environments may be barely visible due to the transformations of the built environment. But, the inhabitation of the city is also ecology of its own. One that is both ordered and not, stable and in constant transition (see Figure 8.6). The flux if habitation results in new ecologies in the lived world.

As I have discussed, place is not an isolated entity, place is one place within many and these multiple places are located in what we name landscape. 'No matter how capacious a landscape may be, it remains a composition of places (Casey 2001, p. 689), full of beginnings and ends. For him the boundary between the embodied

Figure 8.6.

place-world, the place of the geographic self, is best understood through the limitations of our view - the horizon, the border, the line where our current location ends. The placed-self can only know place within the dimensions of the body, the boundary line of the horizon is the marker of the end of the place-world of the geographic self. It marks the line between my embodied self in place and the beginning of something and somewhere else.

> *A landscape is nothing if not expansive. Where the lived body of the geographical self characteristically draws in the place-world around it, ingesting it in the schematized bodily behavior and lingering body memories, the landscape draws out the same place-world, sometimes to its utmost limit.*
>
> Casey, p. 690

Typically we may think of a horizon as the line in the distance that marks the end of our vision. In the countryside this may be kilometers away, in the city the border may be just a few blocks, or drawn by the presence of a building that marks the end of our landscape. These are the markers of the horizon as understood in physical space, however we may also speak of the horizon of our knowing and perceiving. In this context the horizon is an intellectual and ephemeral boundary line of place. I began this text with an account of my technological boundaries and how these formed a barrier between Montréal, and me and then there have been the horizon lines of familiarity, or residency and habatitude. The horizon of place, takes many forms, and like all the elements of place is in a constant state of evolution. For with every step we take, the horizon line shifts, and with that the extent of our knowing and our making of ourselves as geographic beings.

8.7 Conclusions

In the *Roaming Montréal* project I set out to find a trace of the data of the city as they exist in the inhabited city of Montréal, and in so doing, I sought to make tangible the abstract facts that I encountered in the database and the guidebooks.

This was my individual attempt as an outsider in an unknown place, to find my way into this city and to make some sense of the practices of habitation of those who live there. By accepting my role as a transient in this place, I adopted a methodology of movement and observation as a means to be the place-member that I was. This involved a mix of methods, from reading, conversing, looking and seeking, and it was through this combination that I was able to start to translate the multiple data forms and to read between the lines between official and vernacular discourses of place, and to then find traces in the landscape of those who live there. Through this process my status as a transient member of society, a tourist passing through, began to change and in line with Casey's proposition of the inter-relational nature of place-making, a new place began to be formed. The living database began to take form. This included both the lived data of the place becoming tangible, and with that my own geographic self being transformed with a trace of Montréal finding its way on to my being.

References

Abrams, J. & Hall, P. (Eds). 2006 *Else/Where Mapping: mapping new cartographies of networks and territories*. University of Minnesota Design Institute, University of Minnesota Press,

Bachelard, G (translated by Stilgoe, R) 1964, The Poetics of Space, Beacon Press, Boston.

Casey, E 2001. 'Between Geography and Philosophy: What Does It Mean to Be in the Place-Word? *Annals of the Association of American Geographers*, 91 (4), 2001, pp. 683-693

de Certeau, M 1984, The Practice of Everyday Life, University of California Press, Berkley.

Lobeck, A. 1993, *Things Maps Don't Tell Us: an adventure into map interpretation,* The university of Chicago Press, Chicago

Mason, J 2002, Researching Your Own Practice, The Discipline of Noticing, Routledge, London and New York.

Relph, E 1976, *Place and Placelessness*, Pion Limited, London.

Sack, R. D 1997, *Homo Geographicus*, John Hopkins University Press, Baltimore.

Tufte. E. R.1997. *Visual Explanations. Images and quantities, evidence and narrative*. Graphics Press. Cheshire, Conneticut.

Vaughan, L. 2009, 'Walking the Line: affectively understanding and communicating the complexity of place,' *Cartographic Journal,* 46(4), pp. 316-322, Maney Publishing, UK. November 2009

Vaughan, L. 2009b, *Laurene Roaming*, August 25 - October 4, 2009, www.laureneroaming.com

Vaughan, L. & Akama, Y., 2009, 'Engaging with Ku~: from abstraction to meaning through the practice of noticing,' in the proceedings of *The Architecture of Phenomenology Conference,* Kyoto, June 26-29, 2009

Chapter 9

A Post-Naturalist Field Kit: Tools For The Embodied Exploration Of Social Ecologies

Sarah Kanouse

Assistant Professor of Intermedia,
University of Iowa, School of Art and Art History

Abstract

For nearly two hundred years, the figure of the naturalist—the enthusiastic observer of birds, soils, insects, plants, and animals—set the bar for dedicated, non-professional scholarship of the non-human world. With his sketchbook, butterfly net, binoculars, and field guides, the naturalist went "into the field" to learn nature's secrets through patient observation. But recent scholarship in the sciences and humanities has revealed that "the field" cannot be considered apart from the human world that shapes and imagines it. Taking its cue from the study of social nature, *A Post-Naturalist Field Kit* is an art project that updates the figure of the naturalist for the exploration of post-natural urban landscapes. The project includes artifacts for exploring environmental issues in the city—from specimen jars to do-it-yourself air quality monitors and lead contamination tests—along with activity cards that refuse to draw lines between social, economic, and environmental issues. Drawing on Fluxus game kits and recent environmental art, "A Post-Naturalist Field Kit" offers tools for the embodied exploration of urban social ecologies. This article describes and contextualizes the project in light of relevant areas of creative practice and geographical thought.

I stretch my arm awkwardly, reaching with a metal tweezers to scratch and scrape hardened ash from the sill of a window bricked over long ago. The highway roars and trembles not twenty meters away and about as far overhead. The ash resists, turned nearly to stone by decades of indirect moisture, freeze-thaw cycles, and drifting pollen from the milkweed, chicory, and Queen Anne's Lace clamoring for a narrow band of sunlight between this former factory and the raised highway. At last a chunk breaks off and skips to the ground. I stoop to retrieve it. It is greyer and grittier than the dust of the unpaved lot. I pop the hardened ash into a specimen jar, scrawl "780 St-Remi" on a scrap of tracing paper, and place the jar and label into the wooden case at my feet.

I'm a post-naturalist, and this wooden case is my field kit. It contains physical and conceptual tools for exploring Montreal's post-natural urban landscape in the working-class neighborhood of Saint-Henri. By "post-natural," I do not only mean that the city has so modified the ecology of the land that talking about nature is meaningless, although that certainly seems true. Rather, by "post-natural" I also mean to suggest that the traditional *concept* of nature is now somehow untenable, something that we can no longer use as a foil for human activity—if indeed we

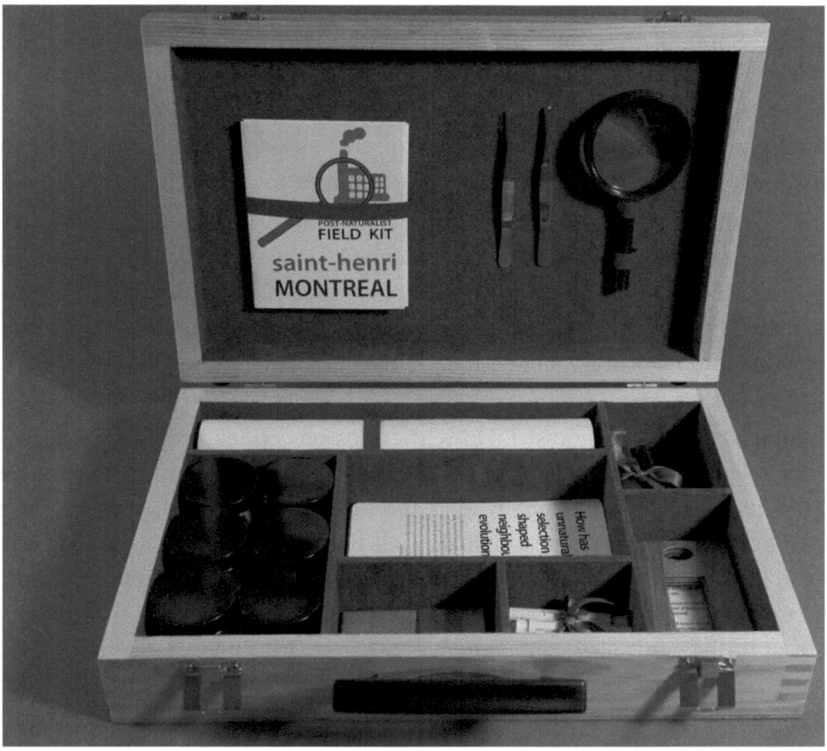

Figure 9.1 "A Post-Naturalist Field Kit for Saint-Henri," 2010.

ever could. In this view, the industrial city is not so much a place where nature has been banished, overcome, or destroyed. Nor are the plants flourishing in the band of sunlight between factory and highway emblems of nature triumphant. Instead, this remote and unglamorous corner of Montreal is a strategic site where the nature-culture dualism—a structuring distinction of modernity—falls apart upon close scrutiny (Latour 1993). What replaces it is a diffuse network of indeterminate interconnection called 'social nature' in which the city's air quality, poverty rates, nesting birds, automobile traffic, native plants, manicured parks, lead-contaminated soil, and condo construction all have a role. "A Post-Naturalist Field Kit" is a toolbox for exploring these interconnections as they play out in the Saint-Henri neighborhood of Montreal.

9.1 A Post-Naturalist Field Kit

My toolbox is housed in a pine case that harkens back to the field kits and butterfly boxes of the classic 18th and 19th century naturalists. Inside, the case is divided into various lined compartments containing tools and items to assist with post-naturalist fieldwork. Some of these are the specimen jars, tweezers, collection envelopes, and magnifying glasses that a naturalist might use, but the post-naturalist is also equipped with wands to test for lead contamination, wooden cards to fashion do-it-yourself particulate air quality monitors, and tracing paper to collect rubbings of the unique textures produced by decaying urban infrastructure. The centerpiece of the kit is a suite of ten, double-sided cards. The front of each card is printed with an open-ended question about the neighborhood, while the reverse side provides three suggested activities—some more practical than others—that might begin to answer the question. The kit is rounded out by a brochure providing background information into the history and current politics in the neighborhood and a map to guide exploration.

Although "A Post-Naturalist Field Kit" treads on serious territory, the instruction cards' questions and suggested activities are oblique and at times tongue-in-cheek. The questions intentionally blur the boundaries between natural, social, and political issues and suggest that they all may be linked. One question asks, "How is the neighbourhood's social ecology linked to other (eco)systems?" while another queries, "How has unnatural selection shaped neighbourhood evolution?" The front of the cards also contain quotations from 19th and 20th century intellectuals too rarely considered together, such as Charles Darwin and Hannah Arendt, Rachel Carson and Walter Benjamin. These quotations amplify the ecological or socio-political connotations of questions and suggest that contemporary conditions have deep historical roots. For example, a question about food sources in the neighborhood is complemented by Karl Marx's critique of industrial agriculture, originally published in *Capital* in 1867. The activities on the reverse side are similarly oblique and contain a mixture of the actually doable and exercises meant to remain thought experiments. For instance, the card about food suggests, "Walk the alleys and parking lots behind restaurants and supermarkets. What and whom do you meet there? How is excess food disposed of? Are the dumpsters locked or unlocked?" By contrast, a card investigating the economic aspects of environmental issues gives the following assignment: "Stand by a highway entrance on a summer's day. Note the number and type of cars that pass in an hour and if their windows are rolled down or up. Express the average Blue Book value of cars with their windows rolled up as a multiple of the average Blue Book value of cars with their windows rolled down." This absurd math problem need not be actually completed to prompt reflection on the ways that affluence shields people from the dirt, noise, and pollution of the roadway.

"A Post-Naturalist Field Kit" draws on the legacy of twentieth century avant-garde movements like Situationism and Fluxus, and well as more recent community-based environmental art. The Situationist dérive was a technique of urban exploration and group research that employed wandering on foot to become attune to the physical, social, and psychological effects of the spaces traversed (Debord 1958). Drawing on artistic antecedents such as Dada and Surrealism and the chance music of John Cage, Fluxus artists used games to promote a spirit of playfulness and encourage a celebration of the everyday. Fluxus games included kits containing objects and instructions, decks of custom-made or modified cards, and modified board games that were often impossible to play with the instructions provided. While my kit treads on more serious territory than Fluxus and is more directive than Situationist dérives, the box is visually influenced by Fluxus games and channels Debordian drifting. It is also influenced by contemporary projects promoting spatial exploration, such as Julian Bleecker and Dawn Lazzi's "Drift Deck" or the Los Angeles Urban Rangers' "Field Guide to the American Road Trip" (Los Angeles Urban Rangers 2006, Bleecker and Lazzi 2008). I drew on the multidisciplinary methods developed at the intersections of art, architecture, and urbanism by groups such as the Center for Urban Pedagogy in New York and X-Field at the Royal Melbourne Institute of Technology in Australia (Center for Urban Pedagogy N.D., X-Field 2010). With this range of inspiration, the content of my cards and use of the objects in the kit remain relatively open-ended; although the suggested activities were developed from research into the neighborhood, they could be pursued in almost any large city in North America. As art, they suggest a playful *method* of urban inquiry, rather than specific content to be learned or conclusions to be drawn.

"A Post-Naturalist Field Kit" promotes a mode of inquiry into urban ecological issues that is attune to the ironies and ambiguities of the post-natural condition and yet utterly sincere in the desire to understand urban environmental

QUESTION THREE

How must the neighbourhood's future contend with its past?

There is a secret agreement between past generations and the present one. Our coming was expected on earth. Like every generation that preceded us, we have been endowed with a weak Messianic power, a power to which the past has a claim. That claim cannot be settled cheaply.

– Walter Benjamin, 1939

A POST-NATURALIST FIELD KIT

Figure 9.2 Sample card for post-naturalist exploration, front.

ACTIVITIES FOR QUESTION THREE

A. Visit one or more of the contaminated sites identified on the map. What do you see, hear, and smell there? Whom do you meet along the way? What sorts of development seem to be in store for the area? Note the types and conditions of plants, animals, and buildings near the site. Ask a fellow pedestrian what they know of contamination at the site.

B. Find signs of the re-use of former industrial facilities, including reclamation by plants and animals, adaptation to new industries, and the management of habitat for the affluent. Which pattern appears to be most common in different parts of the neighbourhood? Who seems to be driving the process?

C. Walk along and beneath the infrastructures of the 19th and 20th centuries—canals, railroads, and highways. What uses and users do you find there? Imagine the infrastructures of the future and how they will be used.

A POST-NATURALIST FIELD KIT

Figure 9.3 Sample card for post-naturalist exploration, back.

issues. Grounding the project to the Saint-Henri area of Montreal is the brochure describing the neighborhood's history, economics, demographics and environmental issues. This background information no doubt colors how the cards are read and affects the activities to be chosen, but it also acknowledges how landscape—even an obviously constructed, urban one—"has the effect of making invisible the operations that made it possible" (Mitchell 1994, 2). In this way, the kit acknowledges that embodied exploration, while necessary, is not in itself sufficient for experiencing multiple dimensions of urban space. Language can provide the contextual knowledge that makes space legible to the explorer. In this way, the cards suggest that a mixture of embodied exploration of physical space, textual research, and the sort of careful observation over time that marked the methods of the classical naturalists may now lead to discoveries about the post-natural city.

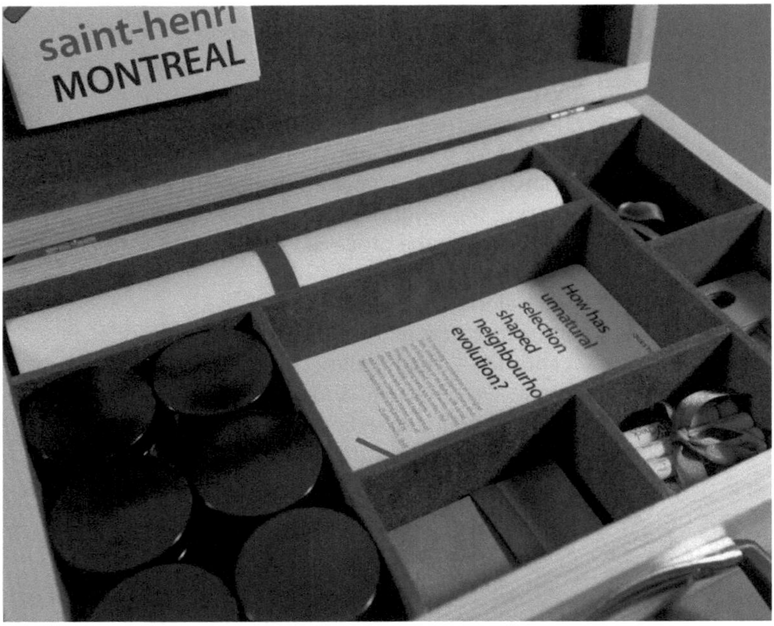

Figure 9.4 "A Post-Naturalist Field Kit for Saint-Henri," 2010 (detail).

On the Post-Natural

Thirty-five years ago, the cultural critic Raymond Williams wrote, "Nature is perhaps the most complex word in the [English] language" (Williams 1983, 219). Since then, nature only got more complicated. The test-tube babies of Williams's time foreshadowed ever-advancing technological capacities to enhance, endanger, modify, and simulate natural processes through industry and technology. Against this backdrop, corporations have shifted from denying the validity of almost all concerns about the environment to embracing the most marketable elements of an environmental agenda as their own (Katz 1998, Rogers 2010). Responses to these changing circumstances in public discourse often involve appeals to nature that are highly contradictory and often politically charged. If everywhere nature is under threat, then it has never been more culturally resonant.

Rejecting the more apocalyptic pronouncements about the end of nature (while taking ecological crisis seriously), scholars from philosophy, cultural studies, geography, and science studies have developed a rich body of inquiry into nature as a social construction. The wide-ranging literature of what has been 'socio-ecology' historicizes nature as a concept; deconstructs the political and ecological consequences of particular notions of nature; explores how nature is produced discursively and materially through human action; and suggests that the contemporary crisis of nature is an opportunity to rethink environmental politics as inextricably linked to science, technology, race, class, gender, capitalism, and democracy (Haraway 1991; Wilson 1991; Latour 1993; Soper 1995; Harvey 1996; Braun and Castree 1998; Whatmore 2002; Latour 2004; Smith 2008; Morton 2010). While a review of this literature is impossible here, these authors take a constructivist approach to nature, influenced by deconstruction and post-structuralism. Although nature has long been viewed as culture's other—the authentic "everything else" to human activity (Soper 1995), sociologist and science studies scholar Bruno Latour historicizes this nature-culture binary by positing it as product of modernity, which depended ideologically on a stark separation between nature (the real), science (the discursive), and culture (the social), even if in practice it produced inescapably hybrid formations that it immediately disavowed (Latour 1993). In this view,'nature' becomes a material as well as discursive formation produced by complex inter-relationships between the material world of organisms, human beliefs and perceptions, and practices that shape, mediate, and, indeed, remake both humans and the non-human world (Castree and Braun 2001).

In many ways, the figure of the naturalist is a product of the structuring nature-culture binary, and he (and the most celebrated naturalists were usually men) can be seen as one of these formations whose actual hybridity was constantly suppressed. With his sketchbook, butterfly net, binoculars, and field guides the naturalist went "into the field" to learn nature's secrets through patient observation. For some, this became a near-spiritual pursuit. "It is blessed to lean fully and trustingly on Nature, to experience, by taking to her a pure heart and unartificial mind, the infinite tenderness and power of her love," John Muir wrote in 1872 (Bade 1924 vol. 1, 325). How the naturalist's tools of observation—the binoculars and field guides most obviously but also practices of observation and Romantic-era notions of nature—affected the observed and the observer were rarely acknowledged. If many naturalists ended up calling for a more holistic ecological view that would consider human beings part of (rather than commanding dominion over) nature, they usually nonetheless saw culture as a corrupting artifice that must be shed to achieve a proper union with the non-human world.

By contrast, socio-ecology proceeds from the observation that "humanity's relationship with nature, in all its permutations, is ineluctable and inherently subversive of the nature-society dualism. From this perspective, human intervention in nature is thus neither "unnatural" nor something to fear or decry. ... [W]hat is at stake is not preserving the last vestiges of the pristine, or protecting the sanctity of the "natural" body, but building critical perspectives that focus attention on how social natures are transformed, by which actors, for whose benefit, and with what social and ecological consequences" (Castree & Braun 1998, 4). In other words, sidestepping the nature-culture binary and embracing the hybrid character of nature-science-culture *assemblages* opens up the possibility of intellectual and political action that is far more complex, inclusive, and justice-oriented than traditional conservation work.

Like many critical positions influenced by post-structuralism, social nature theory has been accused of destructive relativism by environmentalists who believe for political and ethical reasons that nature must be a stable, truthful category of analysis independent of human needs and desires (Soulé and Lease 1995, Deluca 2007). However, it takes seriously questions of social justice, listening carefully to the experience of activist-thinkers in the environmental justice movement for whom romantic and racialized ideas about nature's purity and isolation long impeded efforts to get ecological issues in urban areas—not to mention indigenous peoples' land rights—taken seriously by white, middle-class conservation organizations (Sandler and Pezzullo 2007). Yet social nature theory is not only about adding environmental health and urban sustainability to the agenda of conservation organizations—a process which is well underway (Deluca 2007). Rather, it attempts to understand how the constitution of human-nature relationships are bound up in the totality of the social order—how they are organized and made productive within capitalist modernity and, crucially, how social, political, and economic power relations create uneven and often highly unjust socio-ecological conditions (Heynen et all 2010). Social nature therefore provides a framework for considering how ecological concerns exceed what are conventionally thought of as 'environmental issues.' In this view, housing policy is as much about ecology as energy usage; the routes homeless people use to move through the city become as important to understand as the migration patterns of endangered birds.

While indebted to the experiences of grassroots movements for some of its observations, social nature theory remains rooted in the academy and relies on sophisticated and sometimes highly theoretical arguments about the nature of capitalism and social power. As such, it is not particularly accessible to non-specialists, even as its central arguments are potentially transformative. What would it take for social nature theory's observations to be grasped by non-theorists, including artists, planners, community members, and even high school students? As an artist, I was less interested in attempting to 'translate' this work into layman's terms than in designing experiences in which people might begin to piece together a connective, justice-oriented environmentalism for themselves. Text could and would have to be part of this process, since many environmental issues are invisible and must be explained. But learning new approaches for looking at and inhabiting the city is also necessary for the development of a more connective ecological consciousness. I wanted to employ social nature theory in "A Post-Naturalist Field Kit" to help others to think about urban ecology in an expansive, systemic, and justice-oriented way.

9.2 Developing the Field Kit

In spring 2010, I was invited to participate in an interdisciplinary workshop on art and cartography to be held at Concordia University in September. The event took a novel approach: provide artists and cartographers with a shared database of information on urban environmental issues from which to develop new projects. I had been in the early stages of conceiving an inquiry game kit for the exploration of social nature and urban environmental issues; the opportunity to use very specific GIS data in my research was an unexpected boon. Over the summer, I began to explore data using my self-taught ArcMap skills. Very quickly, the Saint-Henri and Point-Saint-Charles neighborhoods in southwest Montreal stood out as areas that had a disproportionate share of contaminated sites and much higher than average instances of cancer, low birth weight, and respiratory mortality. Some strategic

Figure 9.5 Cracks on the façade of highway-adjacent building in Saint-Henri.

Figure 9.6 Turcot Interchange on the Ville-Marie Autoroute in Saint-Henri, Montreal.

Figure 9.7 McGill University multi-billion dollar hospital complex under construction between working-class Saint-Henri and affluent Westmount.

use of Google revealed that Saint-Henri was facing a major highway reconstruction that threatened to obliterate dozens of homes in this close-knit, working class community. At the same time, gentrification was a growing concern, as was the impact on the area of McGill's hospital mega-project. This trio of issues had made the neighborhood a favorite topic of study for socially-engaged urban planning classes, and a wealth of information about the neighborhood's physical characteristics, demographics, environmental problems, and history was therefore available online. Without leaving my desk in Iowa, I could begin to construct a field kit about Saint-Henri.

Of course, if my aim was to create a toolkit for the embodied and social exploration of urban ecologies, then relying exclusively on data and text sources was inappropriate. After creating a prototype of two cards to offer as an example, I began writing and calling people who work or live in the neighborhood. Some of these were cold calls; others were referrals from artist-activist friends with connections to Montreal. I asked these contacts to share their perceptions of the area's environmental and social issues, and I was impressed by the generosity with which people shared their time, experiences, and opinions. Most respondents referred me to someone else, and I began to get a picture of the various community organizations and citizens' committees. The highway and its proposed reconstruction were immediately and universally identified as the most pressing environmental issues facing the community, and the area's health problems, dangerous traffic patterns, and physical and economic isolation from the rest of the city were all described as related to the highway. Other concerns included a lack of green space, an urban heat island effect, inadequate grocery stores, and dilapidated housing stock. I began to create general activity cards inspired by these responses, my Internet research and the GIS data, and I constructed the box and selected the tools for "A Post-Naturalist Field Kit."

I arrived five days early for the workshop in order to spend many hours walking Saint-Henri and meeting community members, planners and activists face-to-face. I spent three days wandering, often alone, practicing on foot the visual and sensory exploration of ordinary landscapes advocated by artist-forebears like the Situationists, as well as landscape historian John Stilgoe (Stilgoe 1998). Although the kit as an object was complete by the time I arrived in Montreal, the cards and the brochure were edited to reflect what I learned on the ground. For example, a longtime resident took me on a tour and pointed out that tree size on residential streets roughly corresponds to the extent of a block's gentrification. The first streets to gentrify were the ones with the best housing—housing that had been provided to the supervisors of the factories that once employed Saint-Henri's working class. The supervisors of a century ago liked trees, and so do the lawyers and architects trickling into the neighborhood today. This observation inspired a suggested activity on the second round of cards. A card originally asking about railroad noise, dirt, and damage was similarly edited to include highways, based on observations and conversations in the neighborhood.

This iterative, consultative approach in developing the field kit was informed by the mixed methods used by contemporary artists working in the overlapping territory variously labeled community-based, critical, littoral, relational or social

Figure 9.8 Condo conversions in former industrial building on eastern edge of Saint-Henri, near the Lachine Canal.

Figure 9.9 Air pollution on window ledge of highway-adjacent building in Saint-Henri.

practice art. The current iteration of "A Post-Naturalist Field Kit" is not a collaborative, community-based project—the short time I spent in Montreal precluded meaningful dialogue over time, and the conception of the piece is mine specifically. However, elements of the discourse on community-based practice influenced my approach to the prototype I produced and have suggested directions for future projects. Curators like Mary Jane Jacob, critics like Grant Kester and artists like

Suzanne Lacy have advocated that artists working with real-world issues collaborate with and be responsive to the concerns of the communities most affected, eliciting transformative aesthetic expression based on locally-rooted knowledge (Jacob 1993, Lacy 1995, Kester 2004). Striking a somewhat less celebratory note, Miwon Kwon and Claire Bishop have separately cautioned against the artist sidestepping thorny questions of authorship and issues of artistic autonomy in order to present 'the community's voice' as if it were a transparent, unified whole. (Kwon 2002, Bishop 2006). Although these figures often disagree—sometimes strenuously—over the meaning of 'art' as a discipline in this emerging area of practice and how or if artists should strive to be socially useful, elements of their critiques of site-specificity, relationality, and collaboration influenced me to seek out the feedback of local residents and organizers while the broad conceptualization of the project remained my own. I cooperated with local individuals and organizations without presuming to present their experience or obscuring my position as an outsider, and I communicated my hope to use future iterations of the project to more concretely and broadly increase awareness of socio-ecological issues in the neighborhood.

Figure 9.10 Workshop participants drifting toward the highway.

The Field Kit in the Field

"A Post-Naturalist Field Kit" saw its first public fieldwork on the concluding day of the Art and Cartography workshop. I brought workshop participants to the Saint-Henri metro station, distributed cards and brochures to groups of four, and directed them to walk about one kilometer to the parking lot of 780 St-Remi – a building slated for destruction under the government's plan for highway reconstruction.

Figure 9.11 Workshop participants with Pierre Brisset at the Turcot Interchange..

Because the cards' activities did not necessarily match the route we were taking or the available time, each group was also asked to pay particular attention to one feature along the route—green space, traffic patterns, smells, and sounds. In the parking lot of 780 St-Remi, workshop participants were guided to the highway overpass by an architect who had worked with the citizens' mobilization against the highway proposal to develop an alternative that emphasized more sustainable transportation practices and improved quality of life in Saint-Henri without demolishing homes. We then walked to a collective garden to learn from a neighborhood resident about grassroots efforts to improve health, quality of life, and the environment in the neighborhood. I demonstrated the field kit, and everyone discussed what we had noticed on the journey from the metro station to the garden.

Workshop participants were overwhelmingly positive in their response to the field activities. After sitting in the twelfth floor of a downtown building for two days, traveling to a neighborhood far off the tourist maps, feeling the early autumn sun, moving our legs, and smelling an ever-changing mix of cooking, marijuana, and car exhaust began to enact the exploration of urban environmental concerns that we'd spent half the week discussing. The sensory and social co-research, anchored by texts to create context and gentle directives to prompt observation, seemed to strike a chord.

The field kit as an object was less successful in the field. Although aesthetically pleasing, it proved cumbersome and somewhat heavy to carry. It had been designed to echo the historical naturalists' tools, but a contemporary audience accustomed to molded plastic and foam seemed reluctant to use or touch so precious a handmade object. I concluded that it was more suited to be a gallery piece, more useful for imaginative than physical post-naturalist expeditions. I hope to revise the kit concept into something better suited to replication and more expedition-ready. The cards themselves have already been exhibited in a gallery, which I hope to be only one site among many where the project might ultimately live.

This field kit was always conceived as a prototype, and I would like to modify it to act as a self-pedagogical tool for high school and college students as well as interested community members. Unfortunately, attempts to continue working with contacts in Saint-Henri have proven difficult. Although the province's plans for the highway reconstruction appeared to be on hold in September, the Quebec government has since unveiled its counter-proposal for the highway. With scarcely a gesture to citizens' and the city's demands for decreased automobile traffic, more public transportation, and no evictions, the province's latest plan would marginally increase traffic to 300,000 vehicles per day, do little to build public transportation in the near future, and cause over a hundred residents to lose their homes (Riga 2010). This announcement has left neighborhood activists and their institutional partners scrambling to respond, and an art project by someone in Iowa quite understandably and appropriately recedes in the order of priorities. While the project remains firmly grounded in the territory of art right now, what I learned from working on it will inform related work that may find meaning in other spaces, as well.

9.3 Conclusion

> *Our work grows from a belief that the power of imagination is central to the practice of democracy, and that the work of governing must engage the dreams and visions of citizens. (The Center for Urban Pedagogy) believes in the legibility of the world around us. What can we learn by investigation? By learning how to investigate, we train ourselves to change what we see.*
>
> Center for Urban Pedagogy

It is easy to doubt what creative projects contribute to neighborhoods and grassroots organizations struggling with economic, environmental, and social justice issues. What is at stake in community-based work is so much more concrete and palpably immediate: did the family lose its home or not? Are more children dying of asthma or fewer? Yet it is a commonplace complaint by community organizations that their mandates, funding sources, and chronic understaffing force issues to be addressed in a piecemeal, often reformist way, rather than with the sort of expansive thinking that tries to understand intersections between complex issues that often shade into philosophical and ethical questions, rather than technical or tactical ones. Really contending with the recognition of a socio-natural world entails a set of perceptual and political transformations that exceed what community groups, NGOs, and governments can do.

"Thinking the ecological thought is difficult," writes literary scholar Timothy Morton. "It involves becoming open, radically open—open forever, without the possibility of closing again. Studying art provides a platform, because the environment is partly a matter of perception. Art forms have something to tell us about the environment, because they can make us question reality. I would like to stay for as long as possible in an open, questioning mode" (Morton, 2010, 8). The transverse strategies of art—the ones that teach us how to investigate and re-imagine the world—become vitally important. Interdisciplinary creative practice may have far more to do with traditional art forms like drawing—which largely remain the foundation of a young artist's education—than they might seem at first glance. Drawing classes are first and foremost lessons in looking, in seeing beyond the expected, conventional, stylized idea of an object so as to view its unique, contingent form. If interdisciplinary art projects like "A Post-Naturalist Field Kit" use language more than imagery to direct and inform the observer, it is because what we hope to see is not visible to the eye alone.

Bibliography

Badé, William Frederic. 1924. *The Life and Letters of John Muir*. Boston and New York: Houghton Mifflin.

Bleecker, Julian and Dawn Lazzi. 2008. *Drift Deck*. http://www.nearfuturelaboratory.com/projects/drift-deck.

Braun, Bruce and Noel Castree, eds. 1998. *Remaking Reality: Nature at the Millenium*. London: Routledge.

Castree, Noel and Bruce Braun, eds. 2001. *Social Nature*. Malden, MA and Oxford: Blackwell.

Center for Urban Pedagogy. N.D. "Who We Are and What We Do." http://www.anothercup-development.org/about.

Debord, Guy. 1958. "Theory of the Dérive." http://www.bopsecrets.org/SI/2.derive.htm.

Deluca, Kevin. 2007. "A Wilderness Environmentalism Manifesto: Contesting the Infinite Self-Absorbtion of Humans." In R. Sandler and P. Pezzullo (eds) *Environmental Justice and Environmentalism*. Cambridge, MA: The MIT Press, 27-55.

Haraway, D. 1991. *Simians, Cyborgs, and Women*. London: Routledge.

Harvey, David. 1996. *Justice, Nature and the Geography of Difference*. Malden, MA: Blackwell.

Heynen, Nik, Maria Kaika, and Eric Swyngedouw. 2010. "Urban Political Ecology." In N. Heynen, M. Kaika, and E. Swyngedouw (eds) *In the Nature of Cities*. London: Routledge, 1-19.

Jacob, Mary Jane. 1993. *Culture in Action*. Seattle: Bay Press.

Katz, Cindi. 1998. "Whose Nature? Whose Culture?" *Remaking Reality: Nature at the Millenium*. Bruce Braun and Noel Castree, eds. London: Routledge, 46-63.

Kester, Grant. 2004. *Conversation Pieces*. Berkeley and Los Angeles: University of

California Press.

Kwon, Miwon. 2002. *One Place After Another*. Cambridge: The MIT Press.

Lacy, Suzanne. 1995. *Mapping the Terrain*. Seattle: Bay Press.

Latour, Bruno. 1993. *We Have Never Been Modern*. Cambridge, MA: Harvard University Press.

____. 2004. *Politics of Nature*. Cambridge, MA: Harvard University Press.

Los Angeles Urban Rangers. 2006. *Field Guide to the American Road Trip* and *Interstate Road Trip Specialist Field Kit*. http://laurbanrangers.org/site/interstate

Mitchell, W.J.T. 1994. *Landscape and Power*. Chicago: The University of Chicago Press.

Morton, Timothy. 2010. *The Ecological Thought*. Cambridge, MA: Harvard University Press.

Riga, Andy. 2010. "New $3B Turcot to be built by 2018." *The Montreal Gazette*. November 9. http://www.montrealgazette.com/news/Turcot+built+2018/3800058/story.html.

Rogers, Heather. 2010. *Green Gone Wrong*. New York: Scribner.

Sandler, Ronald D. and Phaedra C. Pezzullo, eds. 2007 *Environmental Justice and Environmentalism*. Cambridge, MA: The MIT Press.

Smith, Neil. 2008. *Uneven Development*. Third Edition. Athens, GA: University of Georgia Press.

Soper, Kate. 1995. *What Is Nature?* Oxford and Cambridge, MA: Blackwell.

Soulè, Michael E. and Gary Lease. 1995. *Reinventing Nature?*. Washington, D.C.: Island Press.

Stilgoe, John. 1998. *Outside Lies Magic*. New York: Walker and Company.

Whatmore, Sarah. 2002. *Hybrid Geographies*. London: SAGE.

Williams, Raymond. 1983. *Keywords*. Second Edition. New York and London: Oxford University Press.

Chapter 10

Finding Place: Mapping as Process ...

Annalise Rees

Visual Artist, Port Adelaide, Australia

10.1 Introduction

Through the methodology of a visual artist's approach to mapping Montreal, conventions and modes of cartographic language and literacy are questioned and investigated in regard to expressing and visualising the world in which we live.

Three flights + one taxi ride + five days + sketchbook + pen = Montreal, this is the problem I posed for myself. In attempt to solve this equation through the production of 'art works' with some eloquence and resonance of place, I required a hands-on direct experience with the city. Explored through the process of drawing on and from site this paper outlines my research, discussing the possibility and plausibility of mapping focused on process rather than as complete definitive document.

Place is considered as somewhere we travel to, something we carry with us and also something of ourselves that we leave behind, a trace, a mark. Drawing is posed as an act of reconciling space with self/body, enacted through the execution of a suite of drawings that formed the basis of my research and interaction with the city.

"... the incidents to be narrated were of a nature so positively marvellous... the public were still not at all disposed to receive it as fable, and several letters were sent... distinctly expressing a conviction to the contrary. I thence concluded that the facts of my narrative would prove of such a nature as to carry with them sufficient evidence of their own authenticity, and that I had consequently little to fear of the score of popular incredulity."

A.G. PYM.

NEW YORK, *July 1838.*[1]

Much like Edgar Allan Poe's character, the south sea adventurer Arthur Gordon Pym describes, the questionable authority placed in the hands of the author in delivering truth and accuracy to the public is indeed a daunting task. Expectation and plausibility can far outweigh the cold hard truth. The "*unavoidable exaggeration to*

[1] Poe, Edgar Allen. 1838. *Arthur Gordon Pym and other tales*. London: RE King & Co Ltd, 7- 8.

which all of us are prone when detailing events which have had powerful influence in exciting the imaginative faculties,"[2] can lead us into much more intriguing and satisfying territory, despite a loose grip on reality. When it comes to the map does personal bias always cloud analytical findings? What can provide some surety to accuracy, some definitive record, or is 'truth' defunct?

10.2 Mapping The Real

For millennia the map has played an integral role in defining and representing the 'real.' Used to relay information, visualize and conceptualize space the map has held an authoritative position for understanding our world. JB Harley describes, "*maps as reciprocal images used to mediate different views of the world.*" Their position as a graphic visual language rich in symbolism, iconography and scientific factual information, places the map in a position of apparent power. From this venerated and authoritative stance it can be easy to forget or overlook the relatively humble beginnings of the act of mapping simply as a means of representing spaces graphically; a line, a mark, colour, shape. In essence these acts fall into the category of visual language, one with which an artist is intimately familiar.

The complexity of maps, despite their humble beginnings should not be underestimated by the simple act of linear demarcation. The cognitive understanding and transfer of information from the three dimensional 'real' into the abstracted two-dimensional visual representation of space is a highly sophisticated activity. The process of articulating space and the physical into the abstract form of document and visual communication is a difficult task, as anyone who has picked up a pencil and tried to draw what they see before them can attest to. This problematic transfer and translation of *lived* to *documented* experience is practiced by cartographers and artists alike. The difficulty lies in the telling of 'truth', in the accuracy of faithfully *re*-presenting the real. So in the map have we found the *"statement so minute and connected as to have the appearance of truth,"*[3] that Pym speaks of?

It is possible to recognise maps not only as scientific records as is so readily accepted, but perhaps more usefully so as social and cultural texts. Harley suggests (quoting Muehrcke) "*to acknowledge that all cartography is an intricate, controlled fiction."*[4] He discounts the authority that is given to 'scientific cartography' claiming that deviation, bias and distortion cannot be avoided. Social and cultural context and the role that perception, politics and power play in the depiction and relay of

[2] Ibid.

[3] Poe, Edgar Allen. 1838. *Arthur Gordon Pym and other tales*. London: RE King & Co Ltd, 7- 8.

[4] Harley, J. B. 2002. *The New Nature of Maps: Essays in the History of Cartography*. Baltimore: John Hopkins University Press, 63. (Muehrcke, P. C. 1978. *Map Use: Reading, Analysis and Interpretation*. Madison: JP Publications)

information should be equally recognised. He warns that accuracy deemed truthful, based purely on a scientific approach is dubious and limited, not to mention socially irresponsible and misleading.[5] We also know the worlds maps depict are much more colourful and varied; regularly 'real,' and derived from the physical, sometimes reliant upon experience, but just as often they are imagined, taken from memory, the metaphysical, or simply contemplated.

The commonality of the map within our daily lives is surprising and unashamedly familiar, yet our difficulty in deciphering and decoding their hidden language is frustratingly apparent and frequently felt. Dependent upon the quality of information gathered, its expression and communication within conventional frameworks, the uncertainty of the map as a definitive document of truth presents great difficulty for both author and audience alike. Misinterpretation and misrepresentation are rife within the world of maps, with the complexities of politics, power and struggle defining and distorting the spaces they record. So is it futile to presume that some truth or accuracy can be gleaned from these documents of apparent deception?

10.3 Mapping Montreal

Sept 2nd, 2010.

And so I find myself in Montreal, Canada faced with the fearsome task of creating such a loaded document. A map, a seemingly unassuming three lettered word, meagre in size and yet mighty in implication. I approach mapping from the perspective of a visual artist, not as a cartographer in the *true* or scientific sense, but as someone interested in the experience of 'being in the world,' perhaps more aligned to the phenomenological philosophies of Merleau-Ponty or Heidegger.[6] My interest is in how drawing can physically occupy the spaces we inhabit. From this standpoint my work generally falls into the gallery or exhibition category, but also ventures out into the public domain and is often specific to a site. I think of drawing in both two and three-dimensional terms, as something, which can occupy space, generate place, as well as being representational of it. With these particular interests in mind I use a wide variety of materials to 'draw' with and upon. Drawing for me is a way of reconciling space with self/body and a means of understanding and locating myself within *my* world. My work explores notions of place and identity and the complex relationships and dynamic between self and environment. Drawing is therefore a

[5] Harley, J. B. 2002. *The New Nature of Maps: Essays in the History of Cartography.* Baltimore: John Hopkins University Press.

[6] Heidegger, M. 1962. (trans. John Macquarrie and Edward Robinson). *Being and Time*. New York: Harper and Row.

Merleau-Ponty, M. *The Primacy of Perception.* ed. J. Edie, Evanston: Northwestern University Press.

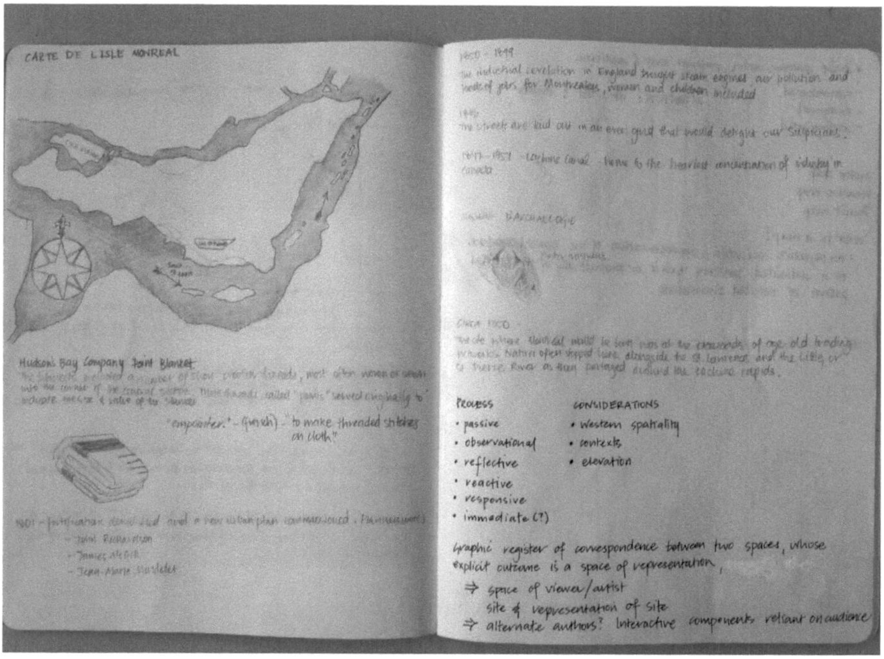

Fig. 10.1 Pages from artist's journal of Montreal.

crucial and central component of describing and decoding these relationships, and in turn my artistic process.

My researching of any location begins with drawings made into a small notebook or journal, along with random jottings, collected scraps and traces of place gathered while walking. Drawing is my primary method of collecting information, that is, recording what I am seeing and experiencing. It is the most direct way that I can impress place upon my memory and consciousness. The time spent drawing and the necessary level of *seeing* involved to record, rather than the brief *glance* of the camera enables me to attain some authority over the images I carry away with me. It is also a reflection of my attentiveness, my level of noticing, concentration and engagement with place. When I look back over old journals it is curious to find that I continue to look at, or perhaps for, similar types of visual information. My eye is caught by similar visual arrangements and relationships, which are in turn recorded repeatedly despite different locations, cities and countries. It is almost as though I carry around an invisible framing device within my consciousness, a radar to hone in on specific arrangements of visual information.

From the journal and initial conventional notions of 'drawing' my investigations then move into more sculptural territory, falling under the broad umbrella of installation based practice. The work often incorporates multiple elements and presents ideas about place and identity through the use of varied materials, and physical

Fig. 10.2 Pages from artist's journal of Montreal.

interaction/occupation of three-dimensional space. The three dimensional work closely refers back to and relies on those initial experiences that I have 'drawn' from life.

Necessary if not crucial for this methodology is my experience of place. Heidegger's *being in the world*, - time to walk, to look, to smell, taste, hear and feel, and Merleau-Ponty's notion of returning to the world which precedes knowledge go some way to explaining this process of encounter. It is the excitement and delight of discovery, which feeds my curiosity and drives the generation of new work. My perception of place is simultaneously created, torn down and re-created through my presence, allowing the construction of meaning within my work. "*All perceiving is also thinking, all reasoning is also intuition, all observation is also invention,*"[7] as written by Rudolf Arnheim describes this essential element of my artistic process and investigation. So perhaps I am a cartographer after all (if I may be so bold as to employ the term within present company)?

Maps are visualised conceptual constructs, visual ideas and descriptions; they convey an understanding of our position within the three-dimensional 'real' world. The complexity of their description and decoding by author/user presents a highly intellectual and sophisticated method of describing ourselves in relation to place/ location in both a physical and socio cultural context. So do my drawings and the journals in which they are contained fall into mapping territory? For me, yes they

[7] Arnheim, R. 1974. *Art and Visual Perception: A psychology of the creative eye.* 2nd edn. Berkeley: University of California Press, 8.

do.

The journals are documents of my encounter with the world and my attempt to visually communicate and express this. They are documents of engagement, *my* record of being in the world, and because of this they carry a truth about *my* experience. James Corner says, "*the power of maps resides in their facticity. The analytical measure of factual objectivity.*"[8] My drawings and recordings are taken from observation, they are objective and visually analytical, but they have equal consideration and acknowledgement of the subjective. My jottings and recordings each day go some way to determining my subjective response to what I saw and felt happening around me. What was worthy of recording? Each response is simultaneously subjective and objective and reliant on the other. Within my exploration both have equal weight and importance.

Broadly across both cartographic and visual art realms the map is created specifically with a viewer in mind. The cartographer presents information to be communicated and interpreted by a viewer and so does the artist. The conventional defining structures that these operate within however, determine two very different types of documents. The map is highly analytical and quantitative; apparently distanced from the author (impartial?), while the work of art; qualitative and highly self-referential. It is this aspect of denying or embracing self-reference I think which stands these two documents apart, but only at a first glance. In the case of the cartographer self-reference is deliberately removed in favour of the 'impartial' truth. As an artist it is specifically *my* experience that drives production and determines the information shared. Defined by the terminology one is objective and the other subjective, however if I return to my earlier explanation, I would argue that my 'maps' are essentially both quantitative and qualitative. In light of writers such as Harley, and Wood it is suggested that the cartographer's map also carries the subjective in equal amounts.

To compare, the cartographer's aim is to objectively present the data, and it is from the accuracy of this process that the map is deemed to have some 'truth.' However as was discussed amongst the workshop group, 'truth' doesn't always equate with effective communication of information. Is truth relative to context, one person's fact is another's fiction, and what about bias? Harley distinguishes between the external and internal power of maps, "*the cartographer has never been an independent artist, craftsman, or technician.*"[9] The power exerted upon maps by politics and the interests of commissioners (and authors) places the cartographer immediately removed from the so-called 'objective' view they are supposed to embody. Further discussion amongst the group posed the interesting question of sacrificing 'truth' for more effective conveyance of information. Sometimes the

[8] Cosgrove, D. ed. 1999. *Mappings*. London: Reaktion Books, 251.

[9] Harley, J. B. 2002. *The New Nature of Maps: Essays in the History of Cartography*. Baltimore: John Hopkins University Press, 63.

facts have to be fudged in order for clarification and effective delivery. Clinging to quantitative facts and truth as being absolute is then perhaps a futile endeavour, and in Harley's view impossible. So I ponder if the effectiveness of maps lies less within their quantitative data and more from their ability to communicate information in some kind of meaningful way or as posed by Wood how the map *"will link its readers to the world it embodies."* [10]

For the purpose of the workshop information was presented in a range of data sets and accessible via the Internet prior to arrival in Montreal. In preparation I printed each of these out and viewed them carefully; columns, lists, figures, and names with some reference to the island of Montreal, apparently. Fortunately for me I was able to remain comfortably distant from the scientific analysis and data, lacking the necessary language to decode the information. With a visual practice predominantly exploring notions of place and identity through the disciplines of drawing and installation, I faced a difficult if not impossible task in identifying with the database supplied. I found the information presented removed from context and without visual reference aside from colour coded charts non relational to my working methods. So I found very little I could respond to. An alternative approach was required, a plan B.

Fig. 10.3 Pages from artist's journal.

[10] Wood, D. 1992. *The Power of Maps*. New York: Guildford Press, 18.

10.4 Drawing Place

Three flights + one taxi ride + five days + journal + pen = Montreal, this is the problem I posed for myself, all thrown in with a dose of jet lag and geo-cultural dislocation. In an attempt to solve this equation through the production of 'art works' with some eloquence and resonance of place, I required a hands-on direct experience with the city. It would be a whirlwind trip with just a few days to encounter and discover. My interpretation of the data would stem from my experience of the city, from walking, sitting, seeing and 'being' in this place. The reference for my map was simultaneously the city and myself. The city became my ultimate point of self-reference, and my map/drawing, the record of this interaction.

Previously I had considered 'place' as somewhere we travel to in the geographic sense, and as something we carry with us from one location to another, a kind of visual and perceptual construct. As I came to understand through this mapping process 'place' might also be something of ourselves that we leave behind; a trace, a mark, such as placing ourselves on or on to the landscape, in this case the city of Montreal. I became both control and catalyst within my experiment with the aim to view objectively while relying upon my subjective experience to quantify the findings.

Drawing as a process of reconciling space with self/body through physical act, inscription and imposition became enacted through the execution of a suite of drawings that formed the basis of my research and interaction with the city. I deliberately chose to work using a black felt tip pen to render some immediacy to the moment of looking. Mistakes and errors in judgement were exposed in the permanency of the ink, erasure and correction forsaken for the truth of being in the moment. The felt tip allowed me to be more concerned with the act of looking than with the 'look' of the drawing. The preciousness of correction possible with the pencil seemed in danger of removing me from the experience of seeing the city, so the felt tip resolved this problem. The difficulty of seeing I hoped would carry with it some integrity, some truth of my experience and so meaning for viewer and author alike.

This path of investigation presented a new framework of thinking in which to consider the act of mapping and drawing through my practice as a visual artist. Using the city as both database, site of representation and interpretation became integral to my mapping process through interaction and observation. From this approach, viewpoint and the authority of the author became important points of consideration. Along with completing my own drawings in situ, I left several notebooks at locations around the city situated in the street, outside community gardens, on park benches and so on.

The notebooks invited passers-by to draw a map explaining where they had travelled from to this point, how they had travelled, their name and age. I fanta-

Fig. 10.4 Notebook left at St Marie Community Garden.

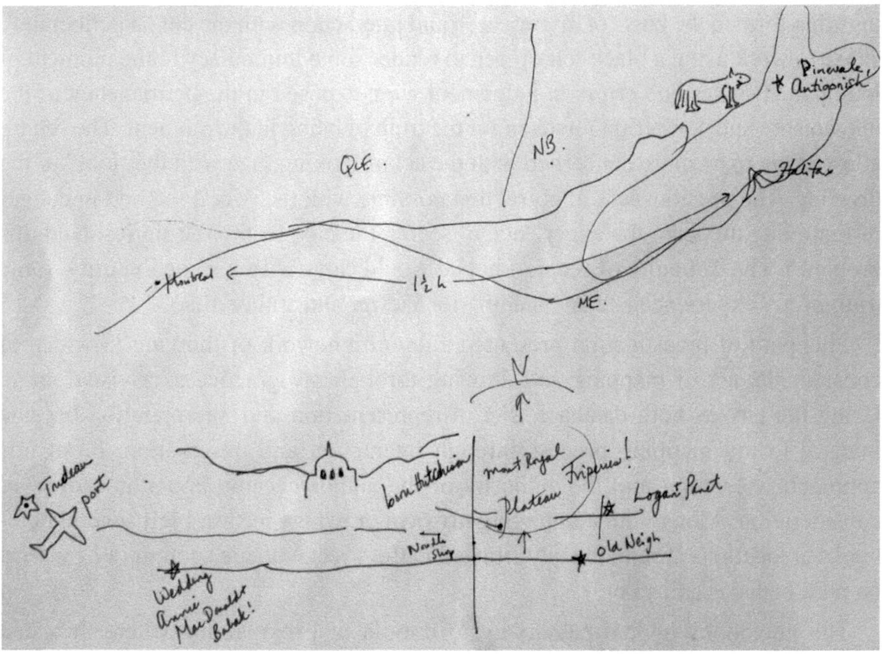

Fig. 10.5 Detail of map drawn by passerby on Rue Panet.

sised about the possibility of accumulating enough hand drawn maps to eventually map the entire island of Montreal – how exciting that city would be to discover! On collecting the maps it then occurred to me that maybe they were reliant on their context, of being within the city rather than removed from it. The maps knew themselves by a described position in relation to their fixity in the site. Would they become meaningless when separated, the data non-referential and so un-readable to the viewer?

10.5 Framing

Back to the group: it seemed the assumption was that the map exists within a predetermined frame, separate from physical context, and most frequently upon the computer screen. While sitting in the workshop on the twelfth floor of Concordia University, the snap shot views looking over the city out of the windows around the room caught my attention. Each view captured beautifully within the window frame presented a contained compositional delight of overlapping geometric shapes and patterns. They were captivating but disjointed views, isolated sections of the city, non relational, disconnected.

Opening up my journal and taking my felt tip I began by drawing the rectangular window frame, placing it comfortably within the confines of the page. With the frame now in place I began sizing up the arrangements of shapes of the view,

Fig. 10.6 Artist's drawn views from Concordia University Building.

carefully visually adjusting proportion and scale to fit into my new window on the page. On each occasion of drawing the view from four of the windows, I found the drawing sprawled out of the designated frame. After the first drawing I especially concentrated on the second to try and make sure the drawing remained within the frame, but again despite my efforts it crept outside, and again in the third and finally the fourth. Using the felt tip prevented me from making corrections or erasing, so each time I had to follow the drawing and work with the mistakes, but each time trying to combat my apparent lack of coordination between what I was seeing and what I was drawing. The drawing seemed to have a mind of its own and suggest where its boundaries lay. It was as though my journey both visually and across the page was to create my own contexts for the city, to think beyond the contained and confined views of what I was seeing. The pen dragging across the surface of the page allowed a new way to traverse, to interact and enact the city.

The problem of the frame presented itself in a similar way while drawing out on the street. This time I used the edges of the page to determine where the drawing would end. Using the felt tip again meant that I would begin somewhere in the middle and work outwards, my eye travelling over the terrain and my pen following as best it could to keep up. In both examples my drawings appeared to be creeping off the page, leading me beyond the edges. I wondered looking back at the drawings what might be going on beyond the edge of the page, in a similar way to the wonder that kept me walking on to see what was around the next bend. It led me to ponder, 'what if the frame becomes fluid and moveable?'

Fig. 10.7 Artist's drawn views from Concordia University Building.

10.6 Locating The Map

The formal notion of the map operates strictly within the frame and safely removed from physical context. It only returns to site through the user, but forever remains distant in its recollection. Stepping out onto the street from the bowels of the metro I popped up map in hand, looking about like a meerkat trying to figure out on which side of the street I was, and therefore if a left or right turn would take me in the desired direction. The map in all its apparent clarity didn't account for metro stations with multiple exits and try as I might to figure out the cardinal points from the sun I wandered off in what I thought was the right direction only to find after ten or so minutes that the left in fact would have been a wiser decision. If only the footpath where I surfaced out onto the street was inscribed with an arrow showing north! I contemplated finding some chalk to help out the next directionally challenged tourist. The uncertainty and mistruths presented in the formal map suggested that an alternative within or inscribed into site could more effectively communicate the required information.

So if the map is to be located within the city, where is it appropriate to draw and what is it appropriate draw on? What are useful points of reference? We know that breadcrumbs didn't work for poor Hansel and Gretel, chalk washes away in the rain and I wasn't too keen on being deported for drawing/graffiti-ing on the street. As I looked for an appropriate location to draw, and maybe draw on, I discovered evidence of the city mapping itself in a number of ways.

Fig. 10.8 Examples of street (mapping) art in Montreal.

10.7 Landmarks And 'Man Marks

Landmarks, and 'man marks, various formal and informal methods of locating oneself were evident all across the city. The usual suspects of more traditional methods, (to be found in most cities across the globe) such as street signage didn't do much in the way of really describing my position in relation to place, apart from the bilingual text. Their unofficial counterparts however displayed the unique character of this French speaking Canadian city and a flavour all its own. There was of course the customary graffiti, synonymous with cities all across the world, not always interesting or creative, more often like discarded cigarette butts – environmentally unfriendly, unsightly and non-descript. With a little closer observation however some wonderful examples of colloquial street culture revealed themselves. Inuit tagging, a script completely foreign to an Australian read more like a drawing than text to unfamiliar eyes, a doodle of presence. Other anglicised versions, with sly humour, political and social comment, expertly executed delivered messages of this specific place and location. Clever, playful, insightful and definitively Quebecois 'notations' were to be found across the city; murals, graffiti, notices about lost cats, dropped notes with unknown names and phone numbers, decorated garden fences, and brightly coloured wool wound around the pole of a parking sign.

All of these unofficial mappings spoke so much more of *this* city, its inhabitants and *real* world action/interaction, and all questioning the authority or value of the autonomous in favour of collective authorship and representation. Considering multiple map users and multiple authors points to online collaborative mapping sites

Fig. 10.9 Evidence of the city mapping itself; architecture and signage.

such as the Open Street Map,[11] and Platial[12]. These operate well within their limited context, but are again removed from actual site and exist purely on the computer screen. What I found most exciting was the implication of the map *re*-presenting location; site and map coexisting in the same given space.

10.8 Documents Of Exchange

Mobile technologies, interactive and collaborative activities such as geo caching, urban projects by groups such as *MapHub*[13], *Blast Theory*[14] and *PVI Collective*[15] are offering alternatives to conventional methods of generation, display and dissemination. GPS and smart phone technology is beginning to allow this to happen, but these devices still operate within a removed context although they can be carried in site. There is something tangible about direct physical interaction/experience (as in the process of drawing) that these wonders of technology are still separating us from. *Grassroots Mapping*[16] with their collaborative, open source approach to participatory mapping present a more site-derived and hands on method, but again their maps are removed from site, but perhaps this is more about the sharing of information and documentation. The *Hand Drawn Map Association*[17] supports a shared web archive of user submitted hand drawn maps and diagrams. As pointed out by a reviewer on the site *"at the time they were drawn, these maps were useful and meant to explain something. But now these saved maps become records of a moment passed."*[18] The map located in-site, re-framing the context as both a physical and cognitive point of reference via direct engagement i.e. the map as landmark and 'man mark with some kind of site specificity and reliance or adherence to location, becomes a much more exciting possibility.

As mentioned earlier it is as though I carry around an invisible framing device within my consciousness, a cognitive mapping of the spaces I encounter. This seems to determine the types of visual information I am attracted to drawing. The built fabric of the city, the architecture and design present innumerable compositions for me to contemplate. The city itself becomes my frame and ultimate point of self-

[11] http://www.openstreetmap.org
[12] http://www.platial.com
[13] http://www.maphub.org
[14] http://www.blasttheory.co.uk
[15] http://www.pvicollective.com
[16] http://www.grassrootsmapping.org
[17] http://www.handmaps.org
[18] Rothman, Julia. 2010. From Here To There. Hand Drawn Map Association website http://handmaps.org/connect.php. Accessed 30 January 2011.

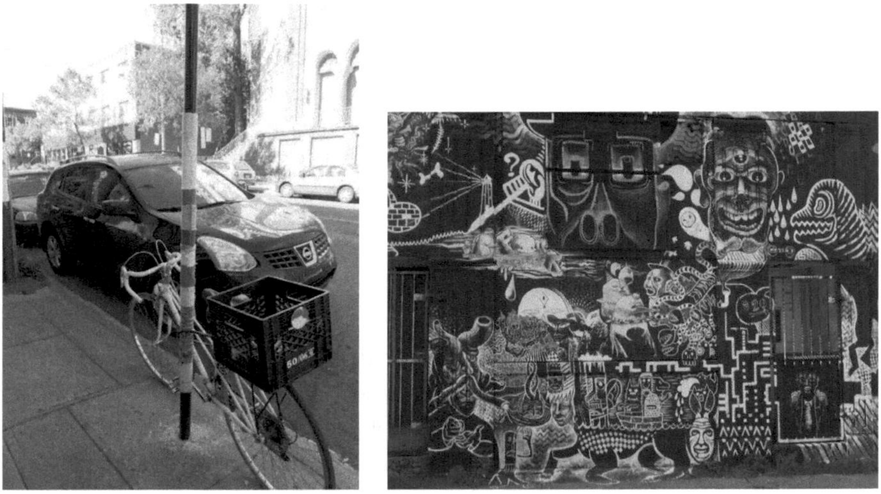

Fig. 10.10 Evidence of the city mapping itself; coloured wool and street art.

reference and difference – the necessary other. The geometry, layering and density of the city fascinate me as the mark of 'man on nature, a way of stabilizing ourselves within the infinite of space and time. Architecture features heavily in my work in both two and three-dimensional form as a signifier of place. Its control and intervention of how we physically encounter and conceptually organise and perceive space presents a means of exploring the relationship between two and three dimensions.

Knowing space, travelling through space and (*re*)generating space all become elements of consideration whilst making mark on surface through the act of drawing. Conventionally drawing is flat and exists on a flat surface (as with mapping), it merely impersonates space or illusionistically reconfigures. What happens when drawing (or mapping) steps outside of these conventional borders of the page and traditional materiality? In my practice drawing is explored and continually pushed and pulled through material exploration into three dimensions. Alternative materials such as wire, light and thread manifest the drawn line into a physical space and sense of being. It is this dynamic, which drives my investigation. When does drawing become sculpture? When does sculpture become drawing and what role does site and location play in relation to the positioning/reading of the work?

Drawing and the map sit precariously within this ambiguous territory. They both construct alternate realities, yet reference the real three-dimensional world. They are often used to describe our experience of place, while frequently being physically and conceptually removed from it. They borrow from and yet simultaneously create their own reality separate from that which they reference. James Corner writes,

"the unfolding agency of mapping is most effective when its capacity for description also sets the conditions for the new eidetic and physical worlds to emerge. Unlike tracings, which propagate redundancies, mappings discover new worlds within past and present

ones; they inaugurate new grounds upon the hidden traces of a living context."[19]

If we consider the abstract geographies that maps construct, why is it so hard to embrace the subjective invention in favour of the analytical and quantitative? Ambiguity and uncertainty concern the cartographers in the group; in it they see a failure of the map as an accurate representation of the data. The discrepancies visually misrepresent and misinterpret. Wood argues that maps construct, not reproduce the world, and states that to take away the frame (or to ignore it) and focus more on the information that is selectively presented in some ethical sense then accuracy is a lesser issue of concern. [20] He gives an interesting example to illustrate this point of a plate from the *Times Atlas* describing the border between Israel and Jordan. Series of dotted and dashed lines speak of various political boundaries with disputed claims to territory. As he points out their location is not uncertain, everyone agrees with their position in respect to latitudinal and longitudinal coordinates. The discrepancy lies however in the ownership and therefore meaning or significance these lines carry for various ethnic groups from the region.[21]

As an alternative if we think about maps as being sites of transfer, as documents of selection and codification rather than as absolute records of the real (as we know this is not the case), then as documents of exchange perhaps they have a greater agency and potential meaning for the user. If we consider maps as sites to trade experience and information, rather than as closed definitive documents there is room for ambiguity.

Interpretation is enacted as the act of making sense and assigning meaning through a process of interaction. As with my own notions of place these can be built up, torn down and constantly reconstructed. Re-locating the map within space and within its location of reference, instead of being removed from place, to allow collaborative authorship via use, potentially opens up a whole new way of thinking about the map. The delight of discovery, of seeing where the road takes you is one of the most satisfying parts of finding oneself within a foreign city. To not know what is around the next corner keeps me walking for hours, simply enjoying the indulgence of seeing and experiencing new things. The map's role is not to replace the real, but to act as an interface, a connection between the real and imagined world. In the case of my inquiry, that interface is enacted through the process of drawing as an ongoing conversation between subjective and objective experience, between the physical and metaphysical.

[19] Corner, James. 1999. The Agency of Mapping. In *Mappings*, ed. Denis Cosgrove, 214. London: Reaktion Books.

[20] Wood, D. 1992. *The Power of Maps*. New York: Guildford Press.

[21] Wood, D. 1992. *The Power of Maps*. New York: Guildford Press, 21.

10.9 Following The Breadcrumbs...

Returning to my journal and the drawings within, they carry some hint of uncertainty and indecision, they speak of a process of inquiry, of deliberate incompleteness. The meandering of the line speaks of the search, the discovery, while the intensity of the ink and its regularity - a particular and attentive level of concentration and engagement. These are perceptive reflections about corporeal looking. Drawings and drawing (noun and verb) are haptic expressions of inscription in and into the world. The stimulus for their enaction cannot be found within spreadsheets of facts and figures. Considering the examples I found within the city as well as my own renderings as tangible 'maps,' these are working documents of inquiry and research. They are a starting point for further investigation and development. It is the act of mapping as an ongoing process that reveals a truth, not in the infinite sense, but more of an intensely intimate and individual representation. The integrity of an intensely personal experience, that hopefully results in some kind of purport for the user.

Ambiguity of experience, uncertainty and discovery consider body as a bridge between being and world, and the necessity of 'lived experience' to construct meaning. Removal from location and *re*-presentation outside the site of reference question the usability and value of the map and the effective conveyance or communication of meaning. It also suggests that multiple viewpoints or perspectives in relation to authorship and audience perhaps may be of more use i.e. the non-autonomous map. If I return to Harley's role of the map as being a reciprocal image used to mediate different views of the world, the map becomes an enacted space rather than a static document. Mediation occurs via its openness and accessibility to multiple user/authors. With the siting of the map within physical context the circumstances

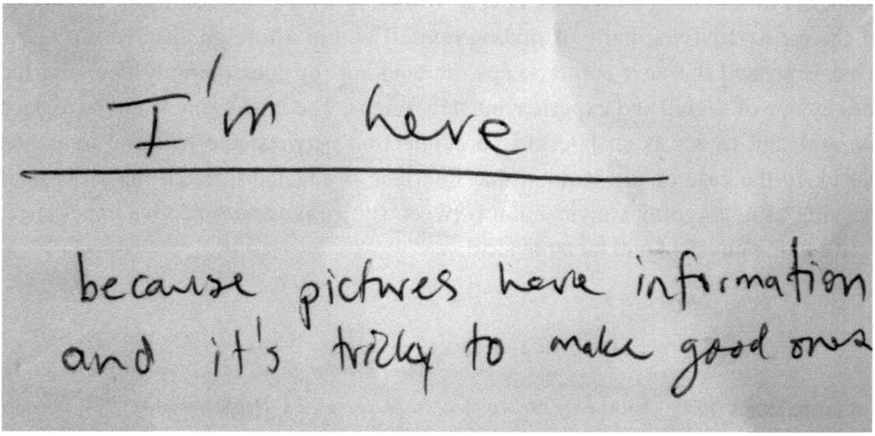

Fig. 10.11 Detail from map drawn by workshop participant in response to on site drawing exercise.

and perceptions of how the map is made, enacted upon and interacted with are evident. Reconstruction of the physical and social settings for the consumption of the map and for it to be meaningful or to hold some truth is not required and therefore accuracy no longer an issue. Instead the map exists simultaneously and symbiotically within the environment from which it is derived; map and site are intravenously obligate.

In respect to my investigation these all became important elements and were evidenced within the city mapping itself, reinforcing the fundamental need for their consideration. My own drawings were a starting point and a necessary cognitive activity for me to experience place and begin pushing my own ideas about how locations can be mapped through the act of drawing. Drawing in its most direct sense will always remain for me the most immediate and truthful way that I can convey and communicate my perception of the world. The humble scratchings of a pen or pencil on paper ignite an infinite range of possibilities through the employ of a broad range of conventional and non-conventional materials. The tools and technologies may change and evolve, however the intrinsic value of marking the world through the act of drawing reveals a truth of experience that carries a quality explicitly unique; that of the human hand which is inevitably connected to the brain. Perception and the value of looking through another's eyes are why artists deem it necessary and ultimately essential to express themselves through the process of artistic investigation.

> *"If we are truly concerned with the social consequences of what happens when we make a map, then we might also decide that cartography is too important to be left entirely to cartographers."* [22]

Through the methodology of this artist's approach to mapping Montreal, conventions and modes of cartographic language and literacy are questioned and investigated in regard to effectively expressing and visualising the world in which we live. The recognition of the equal value of both subjective and objective experience is to be evidenced in the very nature of a workshop of this kind focusing on art and cartography as worthy collaborators. The possibility and plausibility of mapping focused on process and dialogue rather than as complete definitive document, such as in an artist's approach should perhaps be afforded the cartographer without their credibility or authentic value of the map being questioned.

By accepting maps as cultural texts we are able to embrace a number of different interpretive possibilities, but at some point you need to know where you are when it's time to go home. Fortunately for me I was able to follow my slippery, albeit stimulating path back from whence I came to the great southern *Terra Incognita*.

All places are essentially unknown. So we delineate, describe and demarcate,

[22] Harley, J. B. 2002. *The New Nature of Maps: Essays in the History of Cartography.* Baltimore: John Hopkins University Press, 203.

but even then the notion of 'place' exists within the relatively flimsy framework of self-reference, between that of self and other. Containment is impossible, futile and misleading. 'Place' is somewhere we travel to, something we carry with us, and something of ourselves we leave behind. So perhaps to consider the map and mapping in the same light will, if just for a moment carry *the appearance of truth.*

> *... I thence concluded that the facts of my narrative would prove of such a nature as to carry with them sufficient evidence of their own authenticity, and that I had consequently little to fear of the score of popular incredulity.*

A.G. PYM.

NEW YORK, *July 1838.*[23]

References

Arnheim, R. 1974. Art and Visual Perception. 2nd edn. Berkeley: University of California Press.

Cosgrove, D. ed. 1999. Mappings. London: Reaktion Books.

Harley, J.B. 2002. The New Nature of Maps: Essays in the History of Cartography. Baltimore: John Hopkins University Press.

Heidegger, M. 1962. (trans. John Macquarrie and Edward Robinson). Being and Time. New York: Harper and Row.

Merleau-Ponty, M. The Primacy of Perception. ed. J. Edie, Evanston: Northwestern University Press.

Poe, Edgar Allan. 1838. Arthur Gordon Pym and other tales. London: RE King & Co Ltd.

Thrower, N. J. W. 1999. Maps & Civilization: cartography in culture and society. 2nd Edition. Chicago: University of Chicago Press.

Wood, D. 1992. The Power of Maps. New York: Guildford Press.

Wood, D. & Fels, J. 2008. The Natures of Maps. London: University of Chicago Press.

Wood, D. 2010. Rethinking the Power of Maps. New York: Guildford Press.

[23] Poe, Edgar Allan. 1838. *Arthur Gordon Pym and other tales*. London: RE King & Co Ltd, 8.

Chapter 11

Observations

Place as Assemblage: Montreal Garden Mapping

Kathy Waghorn

School of Architecture and Planning, The University of Auckland, New Zealand

Abstract

This paper borrows Manual De Landa's assemblage theory, an elastic schema of society, to consider places as dynamic and extroverted entities, made up of spatial, social and material component parts. Then follows a discussion on the potential of mapping as an operational practice championed in Landscape Urbanism to harness the dynamic qualities of place for making design. Lastly, a description and discussion of Montreal Garden Mapping, a project for the 2010 Arts and Cartography Workshop: Mapping Environmental Issues in the City is given. In this project a landscape urbanist approach to mapping was engaged to map the municipal gardens of Montreal against De Landa's assemblage schema, in order to unfold some of the spatial, social and material component parts and contingent relations of the Montreal place assemblage.

11.1 Introduction

In 2010, as part of the Arts and Cartography Workshop: Mapping Environmental Issues in the City, I bicycled around the community gardens of Ville Marie, an older suburb of Montreal. I did this because I wanted to know about Montreal, not as a tourist, or not only as a tourist, but, as an architect who is interested in the potentialities of marginal urban places, I wanted to see how these places in that city were inhabited. I wanted to know if urban gardening in the city would tell me something about the place that I could not see another way. This Montreal Garden Mapping project is just one small part of a broader investigation of place – an investigation of what places are, how we know them and how we know of them.

It is difficult to pin down a concrete concept of place. Casey, introducing his history of place in philosophy says, "although places are not things in any usual (e.g., material) sense, they are *some* kind of entity or occasion: they are not nothing" (Casey 1998, p. 3, original italics). The paper opens with a discussion of place, considering places as *assemblages*, a nexus of spatial and social components, of material things and expressions, all contingently and temporally located. Next, the capacity of mapping to not only document a given place but to "unfold" the potentialities of places will be put forward. Finally the project, to map the municipal community vegetable gardens of Ville Marie, will be described and framed in relation to the earlier two discussions, of place as assemblage and mapping as a means to convey the dynamic qualities of place.

11.2 Place

In an attempt to identify and re-habilitate a philosophical discussion of place, Casey traces its importance as a concept through western philosophy from Plato to the present. He points to the difficulty yet sustainability of the concept where, even if it is by no means univocal, "'place' is not an incoherent concept that falls apart on close analysis, nor is it flawed in some fundamental manner, easily reducible to some other term, or merely trivial in its consequences" (Casey 1998, p. xii). Place is not nothing, it emerges from the convergence of material, spatial and social factors.

Consider this description of a place -

The city does not consist of this but of the relationships between the measurements of its space and the events of its past: the height of a lamppost and the distance from the ground of a hanged usurper's swaying feet; the line strung from the lamppost to the railing opposite and the festoons that decorate the course of the queen's nuptial procession; the height of that railing and the leap of the adulterer who climbed over it at dawn; the tilt of a guttering and a cat's progress along it as he slips into the same window; the firing range of a gunboat which has suddenly appeared beyond the cape and the bomb that destroys the guttering; the rips in the fish net and the three old men seated on the dock mending nets and telling each other for the hundredth time the story of the gunboat of the usurper, who some say was the queen's illegitimate son, abandoned in his swaddling clothes there on the dock.

Calvino 1972, p. 10

As Italo Calvino so elegantly illustrates place is at once "experienced, structured and discursively constructed" (Dovey 2010, p. 13). That is, we know places, and know of places,

- through our direct experience of them
- through the signs and symbols used in them and to articulate or represent them (consider street name signs, exit and entry signs etc., but also the iconography of places such as the Eiffel Tower for Paris or the London tube map)
- and through the meanings engendered by them. These meanings may vary from the general, such as the common understanding of home meaning a place of security and privacy, to the specific, for example "Chernobyl" as a place is inseparable from the events that occurred there. As Wortham-Galvin notes "a place has truth based not just on the facts of its existence, but also on the things believed to be true about it" (Wortham-Galvin 2008, p. 32).

Approached through diverse disciplines the concept place ranges between two extremes. For some place is bounded and relates to 'genius loci' to 'being', and to stasis. Others reject this essentialist notion of place, and assert place as a dynamic and contested concept about 'becoming'. The geographer Doreen Massey suggests that places are conceptualized as processes (Massey 1991, p. 29), they can, she says, be "imagined as articulated moments in networks of social relations and understandings" (Massey 1991, p. 28). It is this dynamic concept of place that was investigated in this project.

11.3 A "Progressive Sense Of Place"

In rejecting a bounded or essentialist version of place Massey (1991, 1993) identifies four attributes of a "progressive" sense of place. Firstly, place is not static. The material and non-material composition of places is in a constant state of flux. Changes may be permanent or part of a recursive cycle; in the material realm buildings may be demolished and replaced, trees may lose their leaves and gain them again, in the non-material realm land title might change ownership or legislation may re-determine the practices that may be carried out in a specific place. As place is partly derived from social relations places must also change as these sets of relations change.

Secondly, for Massey, places are not enclosures with set boundaries. She acknowledges that boundaries are useful for some studies in relation to specific places, but the boundary as a general concept is not useful for a conceptualization of place itself. All places can be linked to others; places are in fact "extroverted", (Massey's term) formed by the constellation or concentrations of such links. It is these links rather than any boundaries that are important to the construction of specific places.

Thirdly, places do not have singular identities. Places are full of internal conflicts with each person in a place holding their own version of it. In relation to this, finally, for Massey a "sense of place" is important, but this specificity of place arises from a place's external relations, rather than from a turn to "historicism", an internal set of relations portraying a singular and seamless history. Massey promotes what she terms a "global sense of place" in which the character or uniqueness of a place is constructed by the links that place has to many places beyond it.

Massey's global sense of place asserts that places are constituted through these external links, links that occur at multiple, diverse scales. She says,

> *Instead then, of thinking of places as areas with boundaries around, they can be imagined as articulated moments in networks of social relations and understandings, but where a large proportion of those relations, experiences and understandings are constructed on a far larger scale than what we happen to define for that moment as the place itself, whether that be a street, or a region or even a continent.*
>
> Massey 1991, p. 28

Manuel De Landa also considers this issue of scale. Re-invigorating a Deleuzian model, De Landa (2006) puts forward *assemblage theory* as an ontological schema for thinking about social entities. Assemblage theory is geared towards avoiding the intellectual habit of privileging either the macro view, looking at 'society as a whole', or the micro view, examining the routines, categories and behaviors that structure individual experience.

Assemblages, for De Landa, are wholes constructed from the interactions between multiple heterogeneous parts. The parts of an assemblage relate contingently, thus an assemblage cannot be reduced to an essential notion (the essence of a single part) because of this incongruity of the parts. The parts constituting the whole of the assemblage can be assembled in different ways and at different scales based not only on their own properties but also on their *capacities*, that is what the parts are capable of when in combination with other parts (in more or less numbers, denser or looser configurations etc). Moreover these component parts, "may be detached from and plugged into a different assemblage in which its interactions are different" (De Landa 2006, p. 10).

Importantly then assemblage theory operates across micro and macro scales, as De Landa explains,

> *Assemblages, being wholes whose properties emerge from the interactions between parts, can be used to model any of these intermediate entities: interpersonal networks and institutional organizations are assemblages of people; social justice movements are assemblages of several networked communities; central governments are assemblages of several organizations; cities are assemblages of people, networks, organizations, as well as a variety of infrastructural components, from buildings and streets to conduits for matter and energy flows; nation states are assemblages of cities, the geographical regions organized by the cities, and the provinces that several such regions forms.*
>
> De Landa 2006, p. 5

Remember that De Landa is proposing an elastic theory of society through assemblage theory, so in his case assemblages are not necessarily material things, although they may include material things as contingent parts. Kim Dovey however specifically, and usefully for this essay, lends assemblage theory to thinking about place. Dovey says that, "all places are assemblages" (Dovey 2010, p. 16), and he takes a street as an illustration:

> *A street is not a thing nor is it just a collection of discreet things. The buildings, trees, cars, sidewalks, goods, people, signs etc. all come together to become the street, but it is the connection between them that makes it an assemblage or a place. It is the relations of building-sidewalk-roadway; the flow of traffic, people and goods; the interconnections of public to private space, and of this street to the city, that make it a 'street' and distinguish it from other place assemblages such as parks, plazas, freeways, shopping malls and market places.*
>
> Dovey 2010, p. 16

A street then is clearly a place assemblage in a De Landian sense.

There is one last aspect of assemblage theory that must be discussed in applying it to a dynamic concept of place. For Massey, as we have seen, places are constituted by exteriority, by the constellation of their external relations and links. In a similar vein De Landa says that assemblages are, "wholes characterized by *relations of exteriority*" (De Landa 2006, p. 10, original italics). Moreover, for De Landa (drawing on Deleuze) this relationship of parts can be arranged in a matrix, along two dimensions or axes. One axis positions the material-expressive formations of the assemblage while the other locates territory between states of stability and instability. This second axis De Landa calls the territorializing – de-territorializing axis. The component parts of the assemblage can be arranged along these axes with their positioning dependant on their capacity, alone or in combination, to act materially or expressively in the assemblage and to stabilize or de-stabilize the assemblage.

To make sense of this let's consider this assemblage matrix in relation to a specific place and continue Dovey's discussion of a street. In a street the material axis may contain the road surface itself, as well as that of the sidewalk, the underground water main and the fire hydrant. The expressive aspects of this assemblage may be the variable flow of traffic, the sounds of the newspaper vendor calling out on the street corner, the twice-daily passing of the walking school bus or the occasional activation of the fire hydrant that floods the street surface while dousing a fire. The stabilizing (or territorializing, to use De Landa's term) components of the street assemblage might be the signal light that slows the traffic entering the place, the street sign that gives a name to the place and the numbers on shop fronts or letter boxes that give an order to the place, arranging it spatially and indicating legal property and title. The de-stabilizing (or de-territorializing) of the place assemblage might occur through a radical act such as the wholesale demolition of the buildings on one side of the street, as might occur through natural disaster or major re-development. Also de-stabilizing of the street assemblage could be a change in governance and access, where a street might become severed from the larger assemblage of the city. Consider for example the radical action of the construction of the Berlin Wall in literally cutting streets in two, more than impeding the flow of people or traffic, in fact binding part of the place to an entirely different spatial and social order and place entity. In this example the street remains recognizable as a street, but it is no longer the same place. As De Landa says, "one and the same assemblage can have components working to stabilize its identity as well as components forcing it to change or even transforming it into a different assemblage" (De Landa 2006, p. 12).

Given this view of place as assemblage, as made up of extroverted parts contingently related, a 'sense of place' is not therefore 'essential', a reduction to a single 'truth' of the place, but is instead produced by the contingencies between the material and expressive, stabilizing and destabilizing component parts of the place assemblage. Dovey acknowledges this resistance to reduction inherent to an assemblage view of place: "to see places as assemblages is to avoid the reduction of place to text, to materiality or to subjective experience". (Dovey 2010, p. 17). Assemblage theory is a theory of place that takes into account the temporal, spatial and social complexity of places. It is a way to think about place that avoids the reductive 'master narrative', so often activated as the 'master plan' in urban design terms.

11.4 Mappings

So how might mapping intersect with this assemblage theory version of place? As we have seen assemblage theory presents places as complex, fluid entities made up of the internal and external relations between material, spatial and social components, which act with different capacities, stabilizing or de-stabilizing the place. It seems therefore a difficult proposition to try to represent such complex and fluctuating entities. Moreover, Massey, De Landa and Dovey all warn against the use of reduction, to 'scale down' or 'zoom in' to some essential notion or part in order to talk about a place.

James Corner is an architect and landscape urbanist who champions mapping as an operational process that has the potential to render visible, "multiple and sometimes disparate field conditions" (Corner 1999, p. 214). Mapping, he says, has the capacity to capture the spatial and the social, it "allows for an understanding of terrain as only the surface expression of a complex and dynamic imbroglio of social and natural processes" (Corner 1999, p. 214). While Corner acknowledges that maps are used as a tool for projecting power knowledge, and are thus intimately implicated in imperialist projects, he is invested in the operational or instrumental potential of *mapping* as a way of uncovering and bringing together the dynamic constitution of a place. For Corner mapping can make a place; mapping he says, "unfolds potential; it re-makes territory over and over again, each time with new and diverse consequences" (Corner 1999, p. 213). Furthermore he makes claims for an inevitable correspondence between the representational techniques of mapping, with processes of survey and data collection and techniques of adjacency such as collage and the techniques of design production. (Amoroso 2010, p. 111).

In Corner's view it is the process of mapping itself, with the actions of recognising, assembling, and making visible that constitutes the territory or place. Considering this in relation to the axial schema of De Landa's assemblage, maps then act to stabilise territories or places, which is why they are so powerful. Corner, echoing the heterogeneous nature of the assemblage, says that,

> *Reality, then, as in concepts such as 'landscape' or 'space' is not something external and 'given' for our apprehension; rather it is constituted or 'formed' through our participation with things: material objects, images, values, cultural codes, places, cognitive schema, events and maps.*
>
> Corner 1999, p. 223

This version of 'reality' is akin to De Landa's material/expressive-stablizing/de-stabilzing assemblage version of place. In Corner's view, the potential of mapping as a creative practice lies in it's inventiveness, a focus on what maps can do in terms of assembling or constituting the perceived 'reality' of a place.

This instrumental approach to mapping has been embraced by landscape urbanists.[1] In searching for a temporal and social as well as a spatial method though which to approach sites for urban and landscape design projects, the landscape urbanists invent and use techniques of mapping to constitute the 'reality' of such sites as dynamic places open to forces and flows. This desire, to find new ways of 'seeing' a place, stems from the recognition of the assemblage nature of place and the inability to apprehend this dynamic quality of place through the usual representational tools of architecture. Stan Allen, an architect and colleague of Corner's notes that,

> *Traditional representations presume stable objects and fixed subjects. But the contemporary city is not reducible to an artefact. The city today is a place where visible and invisible streams of information, capital and subjects interact in complex formations. They form a dispersed field, a network of flows. In order to describe or to intervene in this new field architects need representational techniques that engage time and change, shifting scales, mobile points of view, and multiple programs.*
>
> Allen 2000, p. 40

Allen's "shifting scales" recalls Massey and De Landa's inclination to avoid

[1] See Waldheim, 2002 and Mostafavi, 2003, also projects by Field Operations such as their Downsview Park Competition entry, Toronto, 2000.

prioritising the macro or micro view of places and social entities.

Mohsen Mostafavi (2003), who wrote a kind of manifesto for landscape urbanism, identifies this shift in similar way to Allen, a shift from what Mostafavi terms "image-based planning processes" to the use of operative mapping methods. In tune with Corner, Mostafavi claims that these operative methods can re-assemble a place. This approach he says, "aims at a search for the hidden pockets of potential, for opportunities in places where previously there were thought to be none. Consequently the urban surface becomes a site of new and unexpected events" (Mostafavi 2003, p. 8).

This approach, to identify and then describe the nuanced relationship between the material, spatial and social orders of and forces acting on a place, and to describe an approach to design that anticipates change over time, has resulted for the landscape urbanists in this emphasis on processes of mapping. Rather than producing traditional architectural drawings, either orthographic (the master plan) or pictorial (the perspective rendering) these types of projects are process driven and the mappings produced act instrumentally. To use De Landa's terms, the landscape urbanist mappings attempt to set down the component parts and contingent relationships of the place assemblage. Corner says, "I am less interested in maps as finished

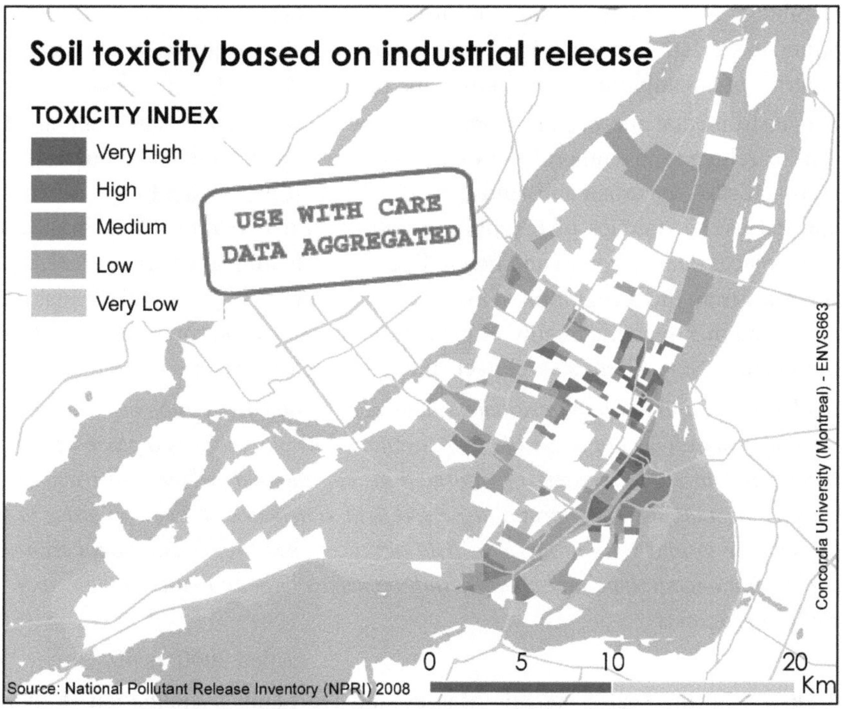

Data compiled and analyzed by Andrew Sanford, Ashvin Ramasamy, Jean-Francois Marsan-Paquin & Undine Thompson

Figure 11.1 Soil toxicity based on industrial release

artifacts than I am in mapping as a creative *activity*" (Corner 1999, p. 217, original italics). It is this potential of mapping that is embraced in the Montreal Garden Mapping project.

11.5 The Montreal Garden Mapping project

The given dataset for the 2010 ICA Arts and Cartography Workshop: Mapping Environmental Issues in the City presented a paradox. While it was the air (quality) and soil (toxicity) of Montreal that was mapped we *see* evidence of neither in the maps themselves. (see Figure 11.1). In order to 'see' the environmental issues of Montreal not as a tourist, and not only through the environmental data, I decided to seek out and locate things that straddle these realms of air and soil, which led me to gardening. Plants, with their roots in the ground and leaves in the air do just this.

In mapping the municipal community gardens of Ville Marie, an older part of the city I hoped to "unfold" (to borrow Corner's term) some of the component parts and contingent relations of the Montreal place assemblage. The project took three phases; firstly the remote gathering of data via web searches, secondly a visit to each garden, photographing their context, use and occupation while recording my own geo-spatial position and physical data, and finally the charting of all found data against De Landa's assemblage model, with its material/expressive – stabilizing/de-stabilizing matrix.

What I first located was a mass of on-line information. Montreal has 8,200 six by three meter garden plots, available to residents in ninety-eight community gardens. I located Montreal city government documents that describe the functioning and administration of the gardens; the procedures and processes for access, the activities permitted in gardens and those that are not, and the policies guiding the requisitioning of land for urban gardening. A report by Canada's Office of Urban Agriculture gave the history of municipal gardening in Montreal, the percentage of population engaged in the scheme, and commented on the economic, social and food safety impacts of the gardens. Québec Government reports addressed the environmental impacts of the gardens and gave detailed accounts of concentrations of soil contaminants, as well as corresponding recommendations for the use and remediation of the gardens. In each of these documents the information was provided by regulatory bodies, with the logos of many agencies adorning the web pages, thus framing gardening as a sanctioned, supported, yet highly regulated activity in the city. Recalling De Landa's assemblage schema, many effective and stabilizing aspects of the place had been located through these documents.

Online versions of maps, produced by many groups (with varying scale and accuracy) helped to spatially locate the gardens. While the satellite images from

Google Maps began to position the gardens in order to plan an itinerary for my visit the seasonal disjunction between the satellite and street view images in Google Maps led to confusion, with the summer/satellite and winter/street view versions of the gardens presenting a vastly different image and sense of place.

For the second phase of the mapping I needed to find an effective way to physically visit each of the twelve gardens of Ville Marie in a single day. For this, another municipally provided service was engaged, the Bixi, a network of rent-as-you-go bicycles. More maps were accumulated as I positioned the twelve gardens in relation to the extensive network of 400 Bixi bike stations.

On a humid late summer's day the next phase of the mapping took place. The nearest Bixi station was located and by swiping a credit card a bike was rented. In order to gain bodily data on the physical experience of visiting the gardens a Garmin Bike GPS was used to record my global position, elevation, speed, heart rate and calories burned. Each garden, its context, contents and occupants were photographed. This second mapping process gave quite a different set of place data. The context of each garden was now recorded, its place among medium density housing, semi-industrial sites and road networks, and its proximity to other services and civic programmes. Some gardens are positioned on marginal pieces of urban land, wedged between other programmes. Some are situated in parks, close to other

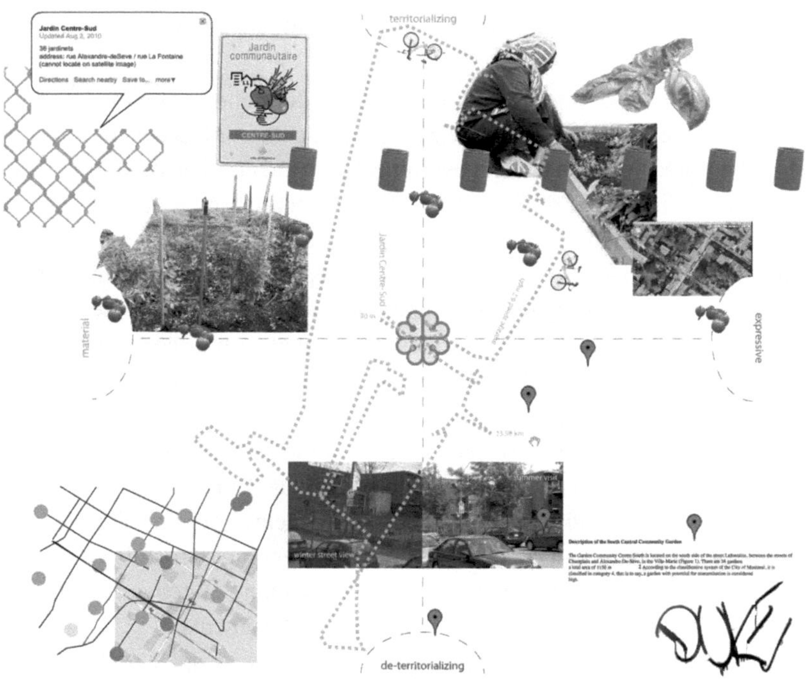

Figure 11.2 Montreal Garden Mapping Project

recreational activities; still others are effectively church gardens. Each garden is wrapped by a high chain link fence and enclosed by a padlocked gate. Each fence carries a sign displaying the garden's name and the community garden logo. Some gardens contain other signage, giving directions for composting or rules for use, all are in French. The garden plots are standardized in size and edging, and identical blue plastic bins were dotted about all of the gardens. Most of the gardens were un-occupied, with only two being tended by sari clad women. This second mapping collated data from across De Landa's assemblage matrix; the material composition of the gardens, the effects of season and position as recorded by the crops, the stabilizing naming of place through signage, the de-stabilizing of the graffiti covering over the same signage and contesting the enclosure of the place.

The final part of the project involved charting the accumulated place data for one garden against the assemblage matrix. De Landa describes assemblages as "wholes whose properties emerge from the *interactions* between parts". (De Landa 2006, p. 5, my italics). This positioning of the component parts against De Landa's matrix proved difficult, as it is only my accumulated knowledge of the place that could allow me to decide how the contingent parts of the place assemblage might be interacting. For example, the sari-clad women tending the garden suggest the existence of new migrant communities as part of this place assemblage. Knowledge of this place assemblage from their experience might allow the contingent parts of this assemblage to then shift in scale, to take into account another place geographically far away and the physical and cultural travel between these two places, invoking Massey's global sense of place. Similarly I can only assume that the graffiti emblazoned sign acted to de-stabilize the control of the garden by the municipal authorities, but this is only an assumption on my part. The soil toxicity data, reporting high levels of contamination, clearly de-stabilizes the place, its functioning as an effective place to grow edible crops, but the official recommendation, a programme and plan for remediation acts against this data to re-stabilize the place. The resulting mapping of the South Central Community Garden can be seen in Figure 11.2.

11.6 Conclusion: The Assemblage Map

The Arts and Cartography workshop allowed for an experiment in applying assemblage theory to a place previously unknown to me. Gathering and charting the data allowed for a complex view of place, both of the gardens themselves and of Montreal as a city. Corner, as we have already seen, advocates for mapping as a creative act, one that can allow places to be seen and registered differently, for their potential to unfold. This mapping project disrupted the tourist view of Montreal that I was more likely to engage, it allowed me to see the complexity of place emerging.

The mapping project promoted a reading of place at multiple scales, macro and micro viewings, from the large scale of city policy, to the small scale of basil, broccoli and soil.

The Montreal Garden Mapping cannot establish a definitive answer in relation to the existence of environmental injustice in Montreal, the question of the workshop, but instead offers a complex view of the urban environment. It charts the city surface, the soil itself, as a multi-scalar dynamic surface, a terrain for an assemblage of heterogeneous parts.

References

Allen, S. (2000) Practice, Architecture, Technique and Representation. London and New York: Routledge.

Amoroso, N. (2010) The Exposed City, Mapping the Urban Invisibles. London and New York: Routledge.

Calvin, I. (1972) Invisible Cities, Orlando Florida: Harcourt.

Casey, E. (1998) The Fate of Place, a Philosophical History, Berkley, Los Angeles and London: University of California Press.

Corner, J. (1999) "The Agency of Mapping: Speculation, Critique and Invention." In Mappings, Dennis Cosgrove (ed.), London: Reaktion, pp 213-52.

De Landa, M. (2006) A New Philosophy of Society: Assemblage Theory and Social Complexity. London and New York, Continuum.

Dovey, K. (2010) Becoming Places: Urbanism, Architecture, Identity, Power. London and New York: Routledge.

Massey, D. (1991) A Global Sense of Place. Marxism Today, pp 24-29.

Massey, D. (1993) Power-Geometry and a Progressive Sense of Place. In Bird, J. (ed.), Mapping the futures: local cultures, global change. London and New York: Routledge, pp 59-69

Mostafavi, M. (2003). Landscapes of Urbanism. In *Landscape Urbanism : A Manual for the Machinic Landscape,* Mohsen Mostafavi and Ciro Najle (eds.), London: Architectural Association, pp 5-9

Waldheim, C. (2002). Landscape Urbanism: A Genealogy. Praxis, 4, pp 10-17.

Wortham-Galvin, B. (2008) Mythologies of Placemaking. Places: Forum of Design for the Public Realm, 20, pp 32-39.

Chapter 12

Observations

Experiments in Correlative Ontography: The Visualization of Environmental and Ontological Injustice

John Calvelli

Alberta College of Art + Design

Abstract

Experiments in Correlative Ontography

This essay discusses a photographic project that was undertaken as a response to a database that was developed to explore issues around environmental injustice in the city of Montreal. As a consequence a new photographic practice was developed, which I name 'correlative ontography'. This practice endeavors to raise questions regarding the visual representation of injustice at both the environmental and ontological level. Through this practice I sought to explore the possibility of representing the ontological 'being' in a photographic image. Potential correlations between geospatially mapped data representing environmental conditions, demographics and health effects, and the photographic images that resulted from this experimental practice of correlative ontography were also examined.

12.1 The Project

A project begins and ends with questions. The questions posed at the beginning of the project determine its future. During the realization of a project, needs are fixed into processes that result in a solution. Once a project is complete, the solution then poses new questions. Some are specific to the nature of the project and its solution; others are common to all projects and will continue on into the future.

Of these solutions, one will ask what remains to be done now that this project is complete. Another asks what kind of future this project will now begin to create. A third remains to be asked:

will this project contribute to more future or will it take future away? (Fry 1999)

This photographic project is itself a response to the database project resulting in, "*Mapping" Environmental Issues in the City: Arts and Cartographic Cross Perspectives*. This database project evolved from a series of questions regarding the relationship between art and cartography and the representation of common data through multiple forms. The database that emerged out of this inquiry resulted in the development of a database and a subsequent workshop. It was the common basis for all the participants' projects, including my own.

I brought to the Mapping Environmental Issues of the City project, a given practice, photography. For two weeks prior to the scheduled workshop, I photographed the city of Montreal, using the database as a reference. This resulted in a series of images that were presented to the workshop participants as propositions questioning the nature of data, our ability to connect to data as a form of visual inquiry, and the relationship of visualized data to the project of creating future..

12.2 The Given Practice

A practice is developed over time. It may have greater or lesser levels of consistency, yet exists within a particular framework of significance. The consistency of my photographic practice was the foundation and the constant element for the purpose of this project.

There have been a number of consistent concerns that have persisted over the duration of my practice: namely, that there is a reality that can be photographed; that this reality can be questioned; that photography can be used aesthetically for the purpose of representation of that reality; and that aesthetic representations may be used to model truth. A creative practice by its very nature responds to changing contexts and concerns. The challenge for the practitioner is to manage the instability that results from this state of change. As a consequence of my participation in this workshop, I separated my given photographic practice from the proposed project of correlative ontography. This allowed the photographic component of the project to perform the role of a constant, whilst introducing the database as a discrete element for the purpose of questions regarding the visual representation of injustice at both the environmental and ontological level.

12.3 What Is Ontography?

Ontography is a visual practice of representing being.

This is at once impossible to do and impossible to avoid. It is impossible to do because all that we can represent is the world, as there is nothing else. It is impossible to avoid, as the world is full of being. For example, a rock may be represented in a photographic image. By representing the rock are we representing being, or are we representing a thing called 'rock'?

In order to live it is necessary for us to respond to the challenges the world places upon us – we are subsumed by the things and actions of the world, and thus forget, being. In a world which is being challenged by it's unsustainability, a world in which being is threatened by non-being, it is important to recognize and be reminded of being. It is my proposition that ontography is a way of recognizing and reminding ourselves of the nature, and the importance of being.

12.4 Correlating Ontography And Injustice

Injustice is the correlating factor in the project of mapping environmental issues in the city and the practice of correlative ontography. Injustice is what exists when some thrive at the expense of others. In the city of Montreal, some inhabitants live in areas where environmental issues increase the likelihood of disease and death, and others don't. The database compiled for the purpose of this workshop became a tool that enabled me to explore the city in order to query if this, indeed, is a form of injustice.

Unsustainability is another, more general form of injustice, one that privileges present over future being. Questions posed through this practice of correlative ontography included the following:

Is the being of the world challenged by unsustainability?

Does the appearance of the world hide its own unsustainability?

Can an image of the world represent unsustainability?

12.5 The Practice Of Correlative Ontography

The results of this project are called Experiments in Correlative Ontography. These experiments attempted to pry the being of the world from the world itself for the purpose of representation. For this attempt, I introduced a constant to the exploration: being-with-a-camera, photographing. This was a real constant, as the click of the shutter is real instantaneity: in the real world, constants can only be relative. I correlate this constant with being-in-the world, photographing. Being is not pictured in this way, but instead being is correlated in a picture. Within the continuum of a practice, the new challenges the constant. When the new impinges on the past (the constant), it needs either to be isolated from the constant as the unique thing it is, or it needs to be incorporated into the constant as part of an integral revision of the constant that makes it more of what it already is.

In a living practice such as the one I bring to this project, the new is always already incorporated as new forms of experience are encountered. The manner in which the new is incorporated is based on preference; in this case it is my preference to incorporate the new reflectively, within a modality of consideration.

This was accomplished by using a map of health data aggregated according to the 30 CLSCs (centre local de services communautaires, or local community service centres) that service the city of Montreal. I used a map generated from the supplied database, "Presence of Diseases Indicative of Environmental Causes" (Caquard et al. 2010), as a means for choosing general locations and neighborhoods to photograph, favoring areas indicating high or low presence of disease. Guided by this map, I chose areas of the city to wander.

The world, already an archive of what has been, became a database whose query was enabled by an eye. As I wandered I photographed rarely, slowly and deliberately, I selected a point-of-view, a place – from which the real could impinge to become an image. A sample was taken as an exposure, shaped and focused through a lens onto a sensor and stored as bits on a card. Care was taken to capture and retain as much information as possible, in the form of detail and tonality. Software and screen were then used to reconstruct the data into an image on screen.

12.6 The Correlated Image

Ten images were selected as visual samples representing ten of the areas I had photographed, and were correlated with a database of environmental, demographic and health data aggregated by CLSC (Romanelli and Roksolana 2010). Rather than choose an image to represent a CLSC according to what was suggested by the aggregated data for that area, I chose an image instead according to criteria that had been established through the constant of my photographic practice.

As an experimental project of correlative ontography, we may ask whether a certain condition of being makes its presence felt in ways that we recognize, correlating with places of documented environmental effects. To achieve this as viewers of images we need to be able to sort out being itself, from being in a specific place, and being in place, from being in an image. Beyond these problems, which might be unsolvable, we need to recognize a particular state of being that could correlate within a specific range of data representing the environmental and health effects as well as the demography of a specific CLSC.

One photograph (Figure 12.3) shows healthy trees and grasses in a riparian area, where there exists a high presence of diseases indicative of environmental causes. Another (Figure 12.8) shows a graffiti- and debris-strewn urban industrial area with a low incidence of similar types of diseases. In other cases, information derived from the database correlate well with a photograph that reinforces our normative visual expectations. Sampling visual strata in the way I proposed and executed in this project, correlating an image with aggregated and geo-specific data, evidently does little to illuminate the possible existence of injustice based on environmental health effects.

There may be many reasons why this is so. The aggregation of the data into CLSCs is likely is too general. Data could be aggregated in a much more limited way, for example, by beginning from the geo-spatial coordinates of the camera position at the time of exposure, and radiating out into the CLSC. Second, the visual strata that was sampled in this study might have little relevance to much of the criteria that was included, such as soil toxicity, number of new immigrants, or people dying of respiratory cancer. One can imagine other kinds of photographic practices, such as photojournalism, that might be applied as a constant to considerably more relevant effect. Finally, as suggested above, a correlative practice based upon the representation of being might be too broad, philosophical or simply impossible to correlate with any given data.

12.7 Injustice, Being and Representation

It was stated at the beginning of this essay that a project begins and ends with questions. The questions that were posed at the beginning of this project include: *Does environmental injustice exist in the city of Montreal*? *How may this be correlated with a visual image*? The photographs and this text are some solutions, existing in material form and posing new questions. Among them is the question of being and its representation; a potentially unsolvable question.

With the publication of this book, the project and workshop, *"Mapping" Environmental Issues in the City: Arts and Cartographic Cross Perspectives*, is presumably finished. The problem of whether there exists injustice, in the form of health effects correlated with location in the city of Montreal, still has no solution. The problem of the existence of injustice and its amelioration, like the problem of being and its representation, are potentially both problems without solution. The insolvability of these two problems poses, at the end of this project, a significant question: *What is the relationship between injustice, being, and its representation?* Injustice has many forms, at different levels of generality and significance. At which levels may we best pose questions that are likely to lead to solutions? Being is the most general of questions. Does its generality promise the most stable of solutions? Or, as the most abstract of questions, does it only reveal the illusory quality of any solution?

These are not, strictly speaking, art or cartographic questions. Yet they may be in-formed by the representation of being.

12.8 Art, Cartography, and Future

Some of the health effects represented in the database for the city of Montreal include respiratory ailments, cardiac conditions and mortality. Of all unjust, environmentally caused health effects, mortality is the gravest. Mortality is the end of a future, a being's future. A point on a map cannot represent the gravity of this event; it is only a place from which we can start to solve a problem.

Today we have questions concerning the future. The ways we have solved problems in our past and present have led to the situation of diminishing future. Being will no longer be; being will be less. We can learn how to represent this threat to being in its gravity; more important is to learn how to represent the starting point, the place from which we can begin to solve its problem.

If these photographs were unable to correlate environmental issues with health effects in the city of Montreal, perhaps that is at least partly due to the need to find this starting point. For we need a one if we are to begin to solve the immense problems of environmental injustice and the possibility of solving the problem of a diminishing future. Perhaps through these photographs we may recognize and remind ourselves of the being that hides itself under a point of a map. It might represent the being that is diminishing, or it might remind us of the being which we wish to preserve.

Figure 12.1 Lot near shoreline with old trees, stumps and rubble pile.

Table 12.1. Disease and demographic data for CLSC Lac-Saint-Louis.

Location	45°25'49.66"N, 73°50'50.74"W
Closest cross streets	Beaconsfield Boulevard and Rue Saint Charles
Median income	$93,862
Low demographic	Low income, single parents, unemployment, adults with no diploma and new immigrants
High demographic	-
Low incidence	Cardiac and cardio-pulmonary disease, lung cancer, underweight
High incidence	-

Figure 12.2 Industrial site from security fence with vegetation and buildings and structures.

Table 12.2. Disease and demographic data for CLSC Dorval-Lachine.

Location	45°26'3.65"N, 73°39'57.50"W
Closest cross streets	Boulevard Saint Joseph and Rue Saint Louis
Median income	$60,338
Low demographic	-
High demographic	Low income, single parents, unemployed and new immigrants
Low incidence	-
High incidence	Cardiac disease and premature delivery

Figure 12.3 Trees, bushes and grasses on the riverbank of Saint Laurent river on île Sainte Helene.

Table 12.3. Disease and demographic data for CLSC Saint-Henri.

Location	45°30'38.79"N, 73°31'52.81"W
Closest cross streets	Chemin Macdonald and Pass du Cosmos
Median income	$45,165
Low demographic	-
High demographic	Low income, single parent and unemployed
Low incidence	-
High incidence	Underweight births, premature delivery, retardation, general and respiratory mortality, cancer and lung cancer

Figure 12.4 Reclaimed urban landfill at the Saint Michel Environmental Complex.

Table 12.4. Disease and demographic data for CLSC Saint Michel.

Location	45°33'44.98"N, 73°37'12.03"W
Closest cross streets	Rue Jean Rivard and 2nd Avenue
Median income	$41,627
Low demographic	-
High demographic	Low income, single parent, unemployed, no diploma, new immigrants
Low incidence	-
High incidence	-

Figure 12.5 Lot with rubble piles and building under renovation.

Table 12.5. Disease and demographic data for CLSC Metro.

Location	45°33'44.98"N, 73°37'12.03"W
Closest cross streets	Boulevard Rene Levesque and Avenue Hope
Median income	$67,765
Low demographic	No diploma
High demographic	Low income, single parent, unemployed, no diploma, new immigrants
Low incidence	Cardiac disease, underweight births, premature deliveries and lung cancer
High incidence	-

Figure 12.6 House with garbage pile in driveway..

Table 12.6. Disease and demographic data for CLSC Rene-Cassin.

Location	45°29'11.79"N, 73°38'14.87"W
Closest cross streets	Place Belfield and Avenue Coolbrook
Median income	$65,458
Low demographic	Low income, no diploma
High demographic	-
Low incidence	Lung cancer
High incidence	Cardiac disease, underweight births, premature deliveries

Figure 12.7 Reclamation site near Vendome station with dirt pile, wildflowers and weeds.

Table 12.7. Disease and demographic data for CLSC NDG-Montreal-Ouest.

Location	45°28'25.31"N, 73°36'11.06"W
Closest cross streets	Avenue de Vendôme and Rue Sainte-Catherine Ouest
Median income	$55,667
Low demographic	No diploma
High demographic	New immigrants
Low incidence	Cardio-pulmonary disease, underweight births, premature delivery and cancer
High incidence	-

Figure 12.8 Loading dock site with graffiti, stencils, tires and debris.

Table 12.8. Disease and demographic data for CLSC Saint-Louis-du-Parc.

Location	45°31'37.11"N, 73°35'45.84"W
Closest cross streets	Avenue de Gaspé and Rue Saint-Viateur East
Median income	$53,426
Low demographic	No diploma
High demographic	-
Low incidence	Cardio-pulmonary disease, underweight births, premature delivery and cancer
High incidence	-

Figure 12.9 Underpass with Cherub-patterned mattress.

Table 12.9. Disease and demographic data for CLSC Des Faubourgs.

Location	45°30'49.94"N, 73°33'1.77"W
Closest cross streets	Rue Notre Dame Est and Rue Saint-André
Median income	$44,102
Low demographic	-
High demographic	Low income, single parent
Low incidence	-
High incidence	Cardio-pulmonary disease, underweight birth, premature delivery, retardation, mortality, respiratory mortality, cancer and lung cancer

Figure 12.10 Side street with grafitti and weeds.

Table 12.10. Disease and demographic data for CLSC Plateau Mont-Royal.

Location	45°31'30.59"N, 73°34'43.82"W
Closest cross streets	Rue Saint-Christophe and Rue Marie Anne
Median income	$57,379
Low demographic	Single parent, unemployed and no diploma
High demographic	-
Low incidence	Cardiac disease, premature delivery, underweight birth and retardation
High incidence	Mortality

References

Caquard S, Cartwright W, Vaughan L "Call." Arts & Cartography Workshop: Mapping Environmental Issues in the City. International Cartographic Association. http://mappingworkshop.wordpress.com. Accessed 23 November 2010

Fry T (1999) A new design philosophy : an introduction to defuturing. UNSW Press, Sydney

Romanelli C and Roksolana B (2010) "Simplified version of the database," Arts & Cartography

Workshop: Mapping Environmental Issues in the City. International Cartographic Association. http://mappingworkshop.wordpress.com/database/. Accessed 27 December 2010

Chapter 13

Observations

Did you Hear the Trains Singing?

Maryclare Foá

University of the Arts London

Abstract

The material, spatial, and social conditions of place can be revealed through ambient sound, while awareness of the affects of sound interacting with place can enhance humankind's understanding of being in place. It was not until I arrived in Montreal that I could know the indigenous sounds of this city: the constant hum of air conditioners and the two-tone sound of the metro trains on entering and leaving their stations. Calling for a refreshed awareness of place, I invited participants of the Art and Cartography workshop to perform a collaborative Montreal Driftsong; a sonic interaction with place. This chapter describes the research undertaken prior to the workshop and the collaborative Driftsong performance of the workshop participants. This includes a reflection on the impact of sound throughout the workshop event, as I pondered the possibilities of sonically mapping place?

13.1 Part I: Research Undertaken Prior To The ICA Montreal Workshop.

'You wish to see, listen. Hearing is a step towards vision'

Saint Bernard of Clairvaux 1090-1153

Since the late 1970's I have been making artwork in the outside environment.[1] By

[1] Inspired by visionary drawing tutor Linda Kitson. Kitson L, (1982) The Falklands War: A Visual Diary, Michell Beazley, London. Kitson was commissioned by the Imperial War Museum in April 1982. http://collections.iwm.org.uk/server/show/ConWebDoc.910

outside, I mean those places that are beyond shelter and are subject to the influences of the weather and the serendipitous or happenstance; such as the passerby who becomes an unplanned witness to an event, or the variations of the natural world and all its inhabitants (animals, birds or plants) who can become unexpected participants. Working in the public domain conditions my practice into what I name Performance Drawing. This is a practice that is always in a state of flux, influenced by, and responding to the unknown witnesses to my actions.

I have explored numerous methods of drawing on paper surfaces in response to place. I have sketched, documented, rendered, marked, and moved through place by foot, car and train. These investigations have led me to suspect that in the process of working in the outside environment an influential interaction occurs within the triangulation of the practitioner the work and the environment.

The writer Merlin Coverley refers to the influence between environment and humankind when he states 'Psychogeography is the behavioural impact of place.' [2] Prior to this project I have undertaken practice based research investigations (at the University of the Arts London), to examine whether an interaction between the outside environment and humankind could be evidenced through drawing.

Realizing that marks on paper do not interact with place, I left the paper surface that had been bridging the gap between place and myself, and began to I mark the ground with chalk and gypsum. However these marks became signs directing the passerby through place, while also (albeit unintentionally), mapping the practitioner (myself) in relation to place.

The unintended trace as evidence of presence, is examined by Jacques Derrida in his study of Emmanuel Levinas's concepts concerning how traces left by those departed, are treasured evidence to those remaining, '.....he will not have been present but he will have left a gift by not disappearing without leaving a trace..'[3] [Derrida 1991]

Derrida's examination of Levinas's ideas of presence, specified the unintentional trace of presence as just that, a trace left-not a mark 'insisted upon.. as a sign'. Yet traces that remain are nonetheless indicators of a previous presence and might be interpreted as directions to follow.While traveling through a place, an animal will follow a track and a person will follow a written direction; and if there is no written direction, a path is then perceived as a primary sign of a tried and tested passage, leading from one place to another.

Bruce Chatwin in his Anatomy of Restlessness tells us 'Children need paths to take bearings on the earth on which they live, as a navigator takes bearings on

---

[2] Coverley M, (2006) Psychogeography. Pocket Essentials, London p 127

[3] Derrida J, (1991) At this very moment in this work here I am. In Berezdivin R (trans) Bernasconi R & Critchley S (eds), Re-Reading Levinas, Indiana University Press, US. p.37.

familiar landmarks,' [4] and even Henry David Thoreau alone in the woods, chose to follow the tracks of the railway line through the wilderness, until 'the bell rings and I must get off the track and let the cars go by.'[5]

Figure 13.1 Children playing with a line drawing life size deck plan (by the author) of the SS Great Britain in Kingstairs Gardens London 2006 m.foá

However these signs in the environment attract attention away from the environment itself; a traveler will concentrate on being directed from one sign to another, rather than being aware of the place they are moving through; and so it follows that a stranger to an unmarked environment will travel with more awareness of the environment than a local. Their survival instincts are fully alert to place as they seek a safe passage through [6]

In an email to road and signage architect Ben Hamilton-Baillie, I described how my drawings on parks were responded to by the passer-by, and added that I suspected that some sort of influential interaction between place and practitioner might exist. Hamilton-Baillie replied, "I'm constantly amazed by the degree to which the outside environment influences our behavior and thoughts, even our perceptions of time passing." And in specific relation to how signs affect place and people in place he added "I'm ever more certain that controls and signals, signs and barriers, ugly road markings and bollards discourage civility, which is the most

[4] Chatwin B (1997) Anatomy Of Restlessness. Picador, London pp.100-106
[5] Bode C (ed) (1982).The Portable Thoreau. Penguin books US, UK. p.374
[6] German Town's traffic Plan: remove Signs, Curbs. By Kyle James N.P.R (National Public Radio) Weekend addition Saturday Jan 19th 2008
[see] http://www.npr.org/templates/story/story.php?storyId=18217318 (24/03/09)

important phenomenon for keeping us safe and connected."[7]

This led me to conclude that my mark drawings on the environment were imposing direction through place, rather than interacting with place. I sought a different method of drawing in response to place.

At this point in my research, during the process of editing performance drawing video documentations, I became aware of how sound interacts with place and that ambient sound can reveal the condition and the material of an environment.

As Susan Delehanty professes, "Ambient sound, or the sound that surrounds us, gives us a sense of our proper bodily location in space." [8] To gain a scientific understanding of this phenomena I met Physicist Dr Eric De Silva (then working for the charity Sense about Science), and learnt that sound is the lowest wave on the gamma scale which while passing through place, is impacted on by the material and spatial qualities of place. Sound resonates, echoes and reflects off and through materials of place, and in this way interacts with place in a three-dimensional expression and measuring (mapping) of its source in relation to place and vice versa. And because the sonic wave is durational, invisible yet trackable, and spatially describes (maps) place, sound, being received by the ear and interpreted by the mind's eye, can be used as a material to draw with, and evidences reciprocal interaction with place.

The connection between spatial and sonic awareness is made evident by sonar technology, and while 'visually impaired humans employ echolocation [estimating]..,'the source of reflected sound in both their ears ...[to] locate themselves.. within a place,' [9, 10, 11], complete lack of sound reflection (as human beings experience when in an anechoic chamber) is spatially disorientating.[12]

Sound's significant capacity to reveal the visual is also explained by writer

[7] see http://www.hamilton-baillie.co.uk/

[8] Delehanty S, (1981) Soundings, Neuberger Museum, SUNY Purchase, , http://www.ubu.com/papers/delehanty.html (Accessed 21/10/07) para 2.

[9] In the 1960s, Caltech Paleoecologist Heinz Lowenstam startled biologists and geologists alike with the discovery that many animals do what conventional science had considered impossible: they manufacture substances such as the iron-containing mineral magnetite within their bodies. Out of Lowenstam's work came the more recent finding that many migratory animals, including birds, bees, and whales, generate magnetite within their bodies and may owe their uncanny homing instincts to the presence of this "internal compass" that allows them to navigate by means of Earth's". magnetic field see http://www.caltech.edu/ (Accessed 23 /11/08)

[10] American Institute of Physics (2008, June 26). Sonar System For The Blind. Science Daily, http://www.sciencedaily.com/releases/2008/06/080625153404.htm (accessed 4/11/08)

[11] Moorehead J (2007).'Seeing with Sound', *Guardian on line*, Jan 27 http://www.guardian.co.uk/lifeandstyle/2007/jan/27/familyandrelationships.family2 (Accessed August 31 08)

[12] Blesser B, Salter L-R (2006) Spaces speak, are you listening? Experiencing aural architecture, M.I.T. Press, US and UK p.18. "..an ideal anechoic chamber is completely silent and entirely "spaceless". I still remember my strange feelings of pressure, discomfort, and disorientation. An aural experience of spacelessness breaches a perceptual boundary ...no longer mask(ing) the sound of a listeners beating heart or flowing blood" Barry Blesser

Elizabeth Finch in her essay The Drawing as Instrument 'understanding that our visual memories are linked to the sonic- we can know something visually by hearing it.' [13]

Another important (often mistakenly regarded as negative) aspect of my research related to spatial orientation, is the fact of my dyslexic condition. As Brian O'Keefe explains in his paper, Dyslexia As A Resource For Design, dyslexia heightens spatial awareness, (and therefore is significantly useful for mapping processes), problem solving and pattern recognition skills (the ability to see relations between topics often perceived as having no connection).

I recognize the connection between observational drawing (endeavoring to fix the relation between one item and another onto a surface) and mapping a place (endeavoring to ascertain the relation between one area and another in place).[14] I also recognize the connection between mapping place materially and sonically.

Having found that sound can be employed as a material to draw with and that sound interacts with place, I came to understand that by vocalising (using my voice box to produce the sound), I became the tool with which to draw through place, and while bridging the gap between place and myself, my vocal sound drawing directly interacts with place.

In relation to cartographic concerns it is useful to qualify the distinctions between the vocal sound drawing process and the mark drawing process, in particular the different materials and their specific effects.

A vocal sound drawing, expressed through the larynx, is activated by breath from the internal body, sent out to the external physical place and then moving through space, leaving no trace. While mark drawings require a tool external to the practitioner and as they delineate a fixed trace; they define, realign and thereby control place. This led me to explore how to sound the outside space, [15] while passing through place, vocalizing, sounding, echoing, and resonating in a repeated rhythmic muted mantra. [16] As Steven Connor tells us in his text Dumbstruck '... there is no other feature [than my voice] whose nature it is thus to move from me to the world,

[13] Finch E (2001) The Drawing As Instrument. in The Drawing Centers Drawing Papers 20, Performance Drawings, Make something in the street and give it away-Alison Knowles, Street Piece, 1962. The Drawing Center, New York, pp.50-54

[14] O'Keefe B (2008) Dyslexia As A Resource For Design http://www.adesignedpath.com/thePath/PHD_Web/611_086_OKeefe.pdf (Accessed 02/02/11)

[15] http://www.nanou.com.au/songlines/ (Accessed 16 May 2009) Chatwin B (1987) The Songlines, Jonathan Cape, UK

[16] Ley lines as well as being traces visible and trackable through place are also invisible magnetic fields, and so are electromagnetic signals. Daniels S (2006) Lines of Sight: Alfred Watkins, Photography and Topography in Early Twentieth-Century Britain. In: TatePapers, http://www.tate.org.uk/research/tateresearch/ (Accessed 05/02/08). Watkins A (1974) The Old Straight Track. Abacus, UK

and to move me into the world.' [17]

While endeavoring to temper my voice to suit each environment, I found that ultimately my voice interacts with place and is shaped through myself, revealing my body and myself. Don Idhe in his text Listening and Voice describes this phenomena as 'the who of voice.' [18]

I have come to call the process of drawing with vocal sound in response to place while passing through place, Driftsinging.

Driftsinging borrows from The Situationist Drift,[19] and Baudelaires Flaneur.[20] Dritsinging also relates to the process of 'Sounding,'[21] the sonic measuring of distance and depth that locates position in place, and 'Echo Location' the examination of place through sonic reflection and refraction, resonance and echo. [22]

Through ultrasound technology our society has understood the process of drawing with sound for around 50 years. [23] Ultrasound produces an image, whereas Driftsinging draws "on the ear,"[24] is experienced through the body, and revealed in the mind's eye. Driftsinging also references Psychogeography '...behavioural impact of urban place,'[25] and Mythogeography [26] where personal associations and history are set in motion and layered into the narrative of place. Driftsongs drawn through place are impacted on by place. In his text A Phenomenology of Voice, Don

[17] Connor S (2000) Dumbstruck: A Cultural History of Ventriloquism, Oxford University Press New York

[18] Idhe D (2007) Listening and Voice; Phenomenologies of Sound. State University of New York Press, US, p.195

[19] Debord G (1958) Theorie de la Dérive. In: International Situationists # 2 1958 Paris December http://www.bopsecrets.org (Accessed 27 March 2009) "a rapid passage through varied ambiences...Dérives involve playful-constructive behaviour and awareness of psychogeographical effects," Guy Debord

[20] Baudelaire C (1964) The Painter of Modern Life and Other Essays, (trans) Mayne J, Phaidon US:, 1964. Constantine Guys was Baudelaire's flâneur, who observed and responded to the flow of urban life.

[21] Sonar originally was an acronym for Sound Navigation and Ranging usually applied to underwater navigation though also used for in air navigation this proc process is also known as Acoustic location or Human Echolocation "By interpreting the sound waves reflected by nearby objects, a person trained to navigate by echolocation.." can identify their location. Wikipedia (accessed 7 June 2009)

[22] Thoreau hearing the timbre of the forest through the sound of the bells. Thoreau H D (1982) The Portable Thoreau. Bode C (ed) Penguin Books, US, UK, p.375

[23] Artist Paul Coldwell alerted me to this fact

[24] Artist Stephen Farthing suggested the idea -Driftsinging draws on the listening ear

[25] Coverley M (2006) Psychogeography. Pocket Essentials.UK

[26] Smith P A (2006) Mis-guided Block Masses (facilitated by Phil Smith) http://mis-guide.blogspot.com/ (accessed 17 May 2009). Artist and member of the Wrights and Sites performance collective Phil Smith's 'Masses' walk (marking the launch of their A Misguide to anywhere book at the ICA 8th April 2006) directed us to bring our personal associations to place. Smith told us that " Mythogeography is the idea that we look in sites that are full of heritage ,(and have) lots of meanings attached to things , (and introduce) a little bit of personal history and related history," which we set in motion through walking , relating and connecting those narratives.

Figure 13.2 The author performing Love: a South London Driftsong 2008 Documentation of performance A. Robinson

Idhe examines sonic interaction when he writes, 'Each thing can be given a voice.' [27] Idhe explains, that things 'sound' in duets, a stick hitting a tree will sound the stick and the tree. Sound, actions objects to sound, and sound is itself actioned upon by the objects it contacts.

I would also offer that sound influences the other's interpretation of place. An animal or human cry imposes a condition of place, while birdsong colors that same place with a very different ambience. In this way an interaction between the environment and vocal sound occurs, and the practitioner is influenced by the materially manipulated sonic responses of place.

13.2 The Collaborative Driftsong Performance

Since 2007 I have been working to develop the Driftsinging method, and have completed a number of solo Driftsongs around South London, Denmark, Australia and America.

In the process of Driftsinging, sonic lines are drawn with the voice responding to and interacting with place. This, sound drawing interacting with place, is a mapping and measuring of spatial dimensions, each action causing a response from the environment and the other (the passerby, witness, and /or spectator). These different

[27] Idhe D (2007) A Phenomenology of Voice. In: Listening and Voice. State University of New York Press, US

responses in turn cause me to modify and reshape my ongoing practice, in this way I have employed 'Action Research' to evaluate my research.

In 2008 I attended a presentation by dowser David Furlong titled London Ley, [28] Furlong described how London is patterned (mapped) with triangles denoting lines of energy connected to sites in and outside the London area. One large triangle has its northern point in Hampstead Heath its Westerly point in Wimbledon and its Easterly point at Greenwich. Furlong explained that the lower line of this triangle carries unbalanced energy. This problematic line runs through Southeast London where I live. I was determined to develop a work in response to this negative condition of place.

David Toop in his book Haunted Weather, describes Japanese composer Takimitsu's duet titled 'Vacalism Ai,' as '.... two voices, one male one female, both repeating the Japanese word ai, (love) in a variety of intonations speeds and pronunciations.' [29]. During a residency at Central Saint Martin's (August 2008), and inspired by Takimitsu's work, I gathered from passers-by on the corner of Southampton way and Theobolds Road, different language translations of the word 'Love.' (Figure 13.3).

My intention; as I performed Love: a South London Driftsong (quietly singing translations of the word Love while walking a section of a London ley line), was to evoke moments of 'Sharawadji' [30] 'Sublime of the everyday, rising out of the bru ha ha sonic muddle.' [31]

To indicate the celebratory intention of the work and to distract the gaze of the other away from my identity, I made a colorful hybrid of a sign, prayer flag, and fascinator.

A colleague pointed out that despite any playful construction, by imposing myself uninvited into the others every day my presence challenges a hierocracy of equilibrium. Rather than meaning to affect a radical challenge, I intend my actions as harmless momentary interventions, temporarily offering a renewed narrative to place. Consequently I realized that during the process of my making various works I had become determined to navigate away from the other (by walking swiftly to avoid interactions), and that I also sought methods which satisfactorily distracted the gaze away from my identity. In short rather than seeking attention I was now

[28] David Furlong presented *Sacred London: its holy wells and hidden alignments* at the Maria Assumpta Centre, Kensington Sq, London 8th May 2008

[29] Toop D (2004) Haunted Weather, Music Silence and Memory. Serpents tail, UK. p.130.

[30] Murray Schafer R (2008) Forward. In: Augoyard J-F & Torgue H (eds.) Sonic Experience, a guide to everyday sounds, Ithaca, McGill-Queen's University Press, Montreal & Kingston, London, pp. xv-xvi

[31] As explained in R. Murray Schafer's text- The word Sharawadji, was brought back to Europe by 17th century travellers to china, to describe an unexpected perception of beauty in the absence of any discernable order. Murry Schafer explains that when this concept is transferred to the urban sound environment "sounds become sublime less by excessiveness than by their implausibility"

Different translations of the word *Love*

(The spelling is written as described by the passer-by who gave the word)

Penda (Swahili) **Lifd** (Dutch) **Khoshmaye** (Kurdish) **Jacal** (Somalia) **Charbliheh** (Norwegian) **Mina** (Pashtu) **Prem** (Hindi) **Habb** (Egyptian) **Szeretlek** (Hungarian) **Kaadhal** (Tamil) **Sayang** (Malaysian) **Fiker** (Ethiopian) **Dueces** (Persian) **Fikr** (Amharic) **Lek** (Hungarian) **Anbu** (Tamil) **Misuri** (Swahili) **Rakkaus** (Finnish) **Love** (English) **Aroha** (Maori) **Chikonde** (Malawi) **Ljubav** (Croatian) **Iubire** (Romanian) **Eu** (Vietnamese) **Ahava** (Hebrew) **Eguono** (Yoruba) **Esso** (Spanish) **Bagnala** (Senegalese) **Ask** (Turkish) **Caru** (Welsh) **Anpu** (Lows) **Adaraya** (Senegalese) **Obicham** (Bulgarian) **Sevgi** (Uzbekistan) **Ai** (Japanese) **Kwagaala** (Luanda) **Mehboob** (Urdu) **Sarang** (Korean) **Mara** (Ugandan) **Hub** (Arabic) **Pyaaar** (Punjabi) **Cint** (Malay) **Ife** (Yoruba) **Mohabhat** (Hindi) **Kocham** (Polish) **Currla** (South India) **Ishiq** (Hindi) **Mhabba** (Maltese) **Houd** (Dutch) **Kaerlighed** (Danish) **Nakupedda** (Swahili) **Prem** (Goodjerati) **Hub**(Arabic) **Dustarm** (Persian) **Agapi** (Greek) **Amor** (Portuguese) **Meile** (Lithuanian) **Amore** (Spanish) **Liefde** (Afrikaans) **Bhalobasha** (Bengali) **Ai** (Mandarin) **Afulaminyai** (Igob) **Uthando** (Ndebele) **Luck** (Thai) **Tumakhay** (Indian) **Liubov** (Russian) **Mahal** (Philippine) **Rudo** (Shona) **Nila** (Czech) **Liebe** (German) **L'amour** (French) **Karlek** (Swedish) **Hob** (Arabic) **Odo** (Ghanaian) **Gra** (Irish) **Bife** (Mali) **Koham** (Polish) **Alskar** (Swedish) **Ima** (Ibibio) **Mikvavxar** (Georgian).

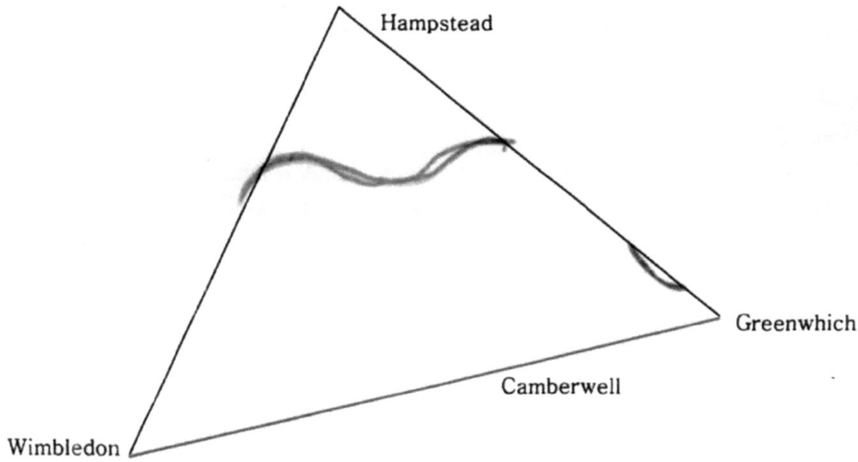

The London Ley Triangle (the blue line indicates the river Thames)

Figure 13.3 Words of Love collected on Southampton Row (written as received) then sung along the South London ley line 18/08/08 m.foá

trying to disappear from the work.[32]

This realization brought me to shift my practice away from myself as a solo practitioner, and towards a collaborative process. I also suspected that my hypothesis (that drawing in sound interacts with place) might be better tested by an action with a broader and multi vocal range. To this end I devised a Multivoice Driftsong score and invited a number of colleagues to take part in this sounding action on Telegraph Hill [33] in South London, on November 2nd 2008.

[32] Artist Hayley Newman disclosed that her motive for making a work in which she tries to disappear was also an attempt to resolve her " …discomfort as a performer and an answer to the problematic of the presentation of the performing ego" H. Newman, "Hook and Eye" in The tingle factor net cast.

[33] In the 1800's Telegraph Hill was a semaphore telegraph station, at the top of the grassy slope

Four friends and three strangers (invited as they passed) took part in the first Multivoiced Driftsong. I directed the seven to stand in a row at the top of the park and listen to the sounds around them. The one at the end of the row listened to and mimicked a sound, then passed it onto the next person and so on until the end of the line. After a couple of rounds I asked the participants to walk as they vocalised the sounds. There were grunts mimicking tennis players, barks mimicking dogs, and swooshing mimicking the wind through trees. Concerned not to take up people's time, I hurried through the process. But participants said they wished there had been longer to listen, and advised I should not be afraid of silence. I resolved to arrange another Driftsong with more participants, and to allow the process to evolve at a natural pace.

Score for a Multivoiced Driftsong **m.foá 02/11/08**

1) In an open space, be the distance of your height away from the other, in front behind and besides
2) Stand silently and still, listen to the sounds around you.
3) Let one begin softly to make vocal sounds responding to the place.
4) The nearest mimic that sound, and pass it on to the next in a flow from one to another, until all have voiced that sound.
5) Again let one begin softly responding to the place in vocal sounds nearest repeating that noise, flowing the sound from one to the next in a wave.
6) Then all begin to walk slowly through place, voicing sound in response to place, renewing the noise as required, and continuing the mimic and flow.
7) Continue this driftsong for five rounds or longer.

Figure 13.4 First score for a Multivoiced Driftsong m.foá

For the second Multi voiced Driftsong, I amended the score clarifying the directions, and selected a central location the history of which connected to my practice.

Bunhill Fields, is commonly known as The Dissenters Graveyard [34] houses the head stones of William Blake, Daniel Defoe and John Bunyon. It could be argued that these three seminal writes, are the fathers of Psychogeography, their metaphysical and physical wanderings of landscape and society, laid the foundations on which later observers, scribed the human condition in relation to the environment, mapping a 'his story' of exploration, wandering and nonconforming.

On March the 1st 2009 over twenty people arrived to take part in the 'Dissenters Driftsong'.I devised a voice warm up session –familiarizing participants with the collective sound of their voices in the open space- and then directed the participants

there is now a sweeping view over millions of houses from Tower Bridge to Battersea Power station and beyond to Hampstead Heath.

34 Originally called Bone Hill Fields, this sight has been burial ground since around the 5th century, and later became the place where nonconformist were laid to rest see http://en.wikipedia.org/wiki/Bunhill_Fields (accessed 28 May 2009)

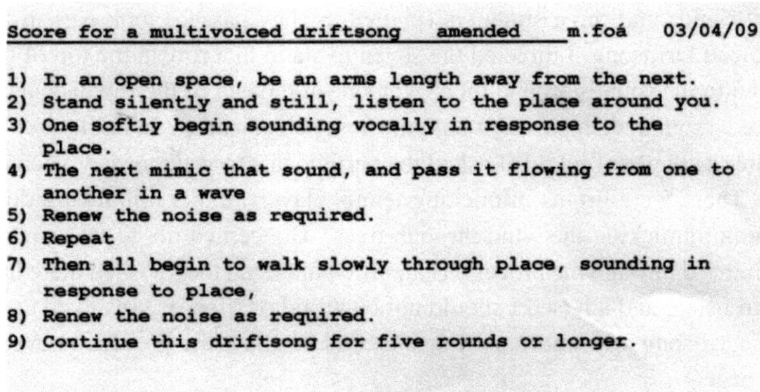

Score for a multivoiced driftsong amended m.foá 03/04/09

1) In an open space, be an arms length away from the next.
2) Stand silently and still, listen to the place around you.
3) One softly begin sounding vocally in response to the place.
4) The next mimic that sound, and pass it flowing from one to another in a wave
5) Renew the noise as required.
6) Repeat
7) Then all begin to walk slowly through place, sounding in response to place,
8) Renew the noise as required.
9) Continue this driftsong for five rounds or longer.

Figure 13.5 Second score for a Multivoiced Driftsong m.foá

to listen to the environment.[35] After practicing vocal mimicking and responding to place, I directed participants to walk together through the graveyard, passing their soundings from one to the next. Later participants told me " there was anarchy, we all felt it's 'my' sound and wanted to stick with it...." and " we ..got the idea we were a drawing, ..we were making a space in terms of the sounds we were mimicking." "you had the John Cage thing happening his listening to the city around you.. "

Despite my camera's limited microphone, during the editing process I found that the Dissenters Driftsong was surprisingly sonically clear. A number of tall buildings around the perimeter of the graveyard had offered resonant acoustics. *I propose that during the Drifsinging action, participants became affected by the sonically receptive place, and the environment,* framing and affecting the sound drawing as it flowed through its space, was also influenced *by resonance temporarily changing the atmospheric coloring of place.*

In April 2010, I took part in the Drawing Out conference at RMIT University and invited the audience to take part in a Melbourne Driftsong. A handful of generous spirited colleagues attended the event, but their evident embarrassment during the process, reminded me that while solo Driftsongs might easily be performed as an instinctive response to place,[36] any performance involving participants who have no vocal confidence or prior experience of public action, requires rigorous planning and tightly choreographed directions.

Following the Drawing Out conference I was invited to submit a proposal for the Art and Cartography workshop in Montreal. At Mappingworkshop.wordpress.com

[35] A robin had discovered his Bunhill Fields 'sweet spot', the particular place within a given area where the conditions most perfectly resonate and enhance a sound. Perched at the end of a leafless branch the bird directed it's song in such a way that all the buildings around the field expanded and echoed the bright little tune into a powerful song.

[36] See Wave for Ned, in which the author sings a standing wave in Ned Kelly's cell http://www.youtube.com/watch/v=id8o5pQdo0s

I read the data base provided for perspective workshop participants,[37] the collection and analysis was part of 'Geographic Information Systems for Environmental Impact Assessment (ENVS663)'. ('The goal of this course was to explore issues related to environmental injustices on the island of Montreal.'[38]) Data concerned the following topics: Socio demographic profiles, Health, and Air Quality. While the database did not contain information on sonic conditions of place, the title of the workshop Art and Cartography Workshop: mapping environmental issues in the city resonated with my research.

Figure 13.6 Participants performing The Dissenters Driftsong in Bunhill Fields London 1/03/09 Documentation of performance S.Stestelo

I employ vocal sound as a method of investigating response to and interaction with place; and through my research I have come to realize that within the vocal sounding process a mapping of some sort takes place, (see page 5 footnote 8 of this paper). I have also come to understand that while vocal sounding facilitates a heightened awareness of environment, sound impacts the environment both negatively and positively; therefore the sonic condition of place is an environmental issue.

In negative form, the sound of vehicles machinery or antisocial behavior, disrupts the lives of habitants, and turns an environment into an inhospitable place. In positive form sound can transform a space or place into an environment of refuge tranquility and rest, sound in place can markedly impact that place for better or worse.

Yet unless we are disturbed by sirens or charmed by birds we are often unaware of the sound of place. Perhaps the chattering of our inner voice accompanied by the various tools that nowadays chirp, sing, buzz and beep at us, all play a part in our tendency to turn a deaf ear to the sound of place.

[37] Compiled by Concordia students in Environmental Assessment Department of Geography Planning and Environment at Concordia university Montreal 2010

[38] http://mappingworkshop.wordpress.com/database/

In as much as S.O.N.A.R locates ships in relation to the sea floor and or any objects between the sea floor and the ship's hull, it can be said that vocal sounding (V.O.N.A.R, vocal navigation and ranging[39]), also locates the person in place. Through a sonic mirroring of reflection and refraction, V.O.N.A.R maps its source in relation to place and visa versa. Environments can be explored and examined through sound; the process reveals material, spatial and often social, conditions of place. In this way awareness of the affects of sound interacting with place, can enhance humankinds understanding of place and also of their being in place.

My understanding of the importance of the sonic activity of place, prompted me to offer to The Art and Cartography workshop, a practical collaborative process (Driftsinging), to facilitate sonic awareness as an additional layer of data that may inform and contribute to the understanding of environmental injustices.

13.3 The Impact Of Sound Throughout ICA's Montreal Workshop Event.

Before travelling to Canada I surfed Street View around Montreal's Concordia University buildings, looking for suitable spaces to Driftsing, but online maps (at present) are deaf and mute, with few visual prompts to suggest the virtual environment's ambient sounds.

I arrived in Montreal city before the start of the conference, and spent an afternoon looking for suitable Driftsong spaces close to Concordia. The air conditioning units were in full voice- sounding a constant drone in duet with my tinnitus.

Each city acquires; mechanically and or organically, a set of indigenous sounds. Some are so familiar that local inhabitants are no longer aware of them. In Melbourne the street crossings, tick until a bell sounds safe passage, in London [40] the public announcement "Mind The Gap" reveals the ancient Underground, (where the curving platform yawns a gap between the train and the platform edge), while the twice evening roar of Concord as she lowered herself towards Heathrow gave the city a more contemporary air (R.I.P). In Montreal together with the noisy air conditioning units, the indigenous sounds are the subway trains singing two notes on their arrival and departure from each station.

My proposal, to investigate the possibility of sonically mapping an area of Montreal in a collaborative performance; required other participants of the workshop to help me map place with sound.

[39] The term V.O.N.A.R has been devised by the author and artist researcher Leon Barker: MiPP (the Motion in Place Project; examining motion capture in archeological fieldworks at Sussex University.)

[40] McCrae D F 1879-1939 'Hark to old London's smothered roar' In: Homesick.

As I realized from previous multivoiced Driftsinging experiences, it is a lot to ask strangers to perform in public. I'm also aware that the process of Driftsinging because there is (currently) no visual evidence of interaction between vocal sound and environment, pushes the boundaries of belief and credibility.

However as I have indicated previously in this paper the underlying intention of my Driftsinging work is to vocally interact with place in order to heighten participants awareness of physically being in place; and the relationship between place and ourselves through vocally sounding place, collaboratively spatially explores place. This multivoiced process highlights the sonic condition of place, thereby contributing to information of environmental conditions.

Driftsinging participants interactively engage with place, while also exploring the concept that sound might be a material that can be used to map place.

Sound is received through the hearing ear and transmitted into the mind; working with echolocation or sonar, the individual comprehends the experience of being physically immersed in sound, and (as with sight impaired people who use stick tapping or tongue clicking) can construct a map of the surrounding space, in the mind's eye.

Through previous experiences of collaborative Driftsong actions I understood there to be a number of problematic social issues to address before attempting to direct a Montreal Driftsong. Firstly I needed to negotiate any embracement, which participants might experience through sounding in public. Secondly in order for the Driftsong to be successful I must proscribe clear directions and a strong structural method defining the action.

I gathered together in brown paper bags, a collection of items (handmade and manufactured) to give to the participants. These Driftsinging handouts served as an incentive to take part in the action, and became souvenirs of the event. Items included background information, directives for participation, a small badge, scores for driftsinging works, music manuscript paper, souvenir cards a cough lozenge and a small box of apple juice.

To dissuade embarrassment, those participants who wanted to disappear, could disguise themselves by wearing the paper bags on their heads, there being eye and mouth holes cut into the sides.

On the 8th September 2010 (8/9/10), almost all those present at the Art and Cartography workshop in Montreal volunteered to take part in the Montreal Driftsong. Having identified two accessible environments close to Concordia University, whose material structure and ambience were distinct from each other. We first visited a small garden on Rue Sherbrooke Oust. One block North and two West of Concordia's SGW campus on Rue Mackay, The Church of St Andrew and Saint Paul have recently renovated a small garden. Situated between the high walls of the church on one side and the high walls of the Montreal Museum of Fine Art on the other, the small green space is accessed by a gravel path. Participants entered the site in line one after the other, already sounding as their footfall crunched over small layered stones.

Arranged in a broad circle I directed a warm up session, soon responded to by the local birds. [41]

Participants were very generous in their collaborative spirit, and positively sounded the small tree filled space. The high walls reflecting resonance and reverberation satisfactorily interacted with our Driftsong.

Next we walked together along Rue Sherbrooke to Rue Mackay, then West through a small alleyway into an air-conditioner filled open parking area, between Rue Guy and Rue Mackay. This urban space humming with the sounds of busy machinery, and full of vehicles, was a strong contrast to the quiet bird song ambience of our first Driftsinging space.

Figure 13.7 Montreal Driftsong participant wearing the paper bag disguise. 08/09/10 still frame from moving documentation of performance m.foá

I had devised a number of different scores that involved a variety of physical responses to sound. In the first participants were asked to gather in a tight bunch and to move towards any sound – this interaction, negating physical boundaries developed into the most successful (and perhaps also most enjoyable) of the different actions. One member being extraordinarily tall towered over the bunch of participants around him, while another playfully exaggerated the lilting, swerving

[41] Young E (2008) Noise Pollution makes city song birds change their tune. In: New Scientist 03/28/2008 http://www.greenchange.org/article.php?id=2376 (accessed 09/05/2009) Urban birds are singing louder to get themselves heard over man and machines.

and jostling of this small tight mass of movement, provoked by ambient sounds. However rather than an investigation into the possibility of sonically mapping place, the Driftsinging actions in Montreal were wholly focused on sonically interacting with place in order to promote individual awareness of the sound of place, and to highlight the importance of sound in relation to the environmental condition of place.

The following day after the Driftsinging event, a number of participants discussed their related sonic experiences. One had agreed with her colleague that they would never in reality stand so close to any other person as they had done in the Driftsinging action, jostling as a tight mass. Yet the next morning on the subway they had found themselves doing that exact same action all over again with a carriage of strangers huddled together; they swayed and swerved towards various sounds.

Figure 13.8 & **Figure 13.9** Art and Cartography workshop participants Driftsinging 08/09/10 still frame from moving documentation of performance m.foá

Another participant Sarah Kanouse had realised the sonic condition of place in her work A Post Naturalist Field Kit Saint-Henri Montreal (see chapter in this volume). Kanouse took us to visit this island of life between roads and rail lines. Saint Henri is a collection of old houses, and a condemned factory building, bordered on one side by a labyrinth of used and disused highways, and the other by a railway line. Standing on a tree lined path planted by local inhabitants, Kanouse observed that on one side of the path crickets and birds can be heard while on the other side of the path the air is filled with the roar of constant traffic.

13.4 Part II: Is It Possible To Sonically Map Place?

Earlier this year at the Drawing Out conference in Melbourne I met ICA vice president Georg Gartner, he talked to me about the impossibility of mapping, "how" he asked "can you pin point which exact tree marks the edge of a forest." His question showed me that mapping is like drawing from life. In drawing there is the untruth, which every draftsperson negotiates whilst rendering a 3d object onto a 2d surface. Unless objects are flat they have no edge- so drawing a line around something is a lie in itself. Yet culturally we have devised a visual language, now accepted by a large proportion of western society, a short hand, read and understood, rendering with marks tone and line, and fixing form onto a surface. In this way drawing is an individual's interpretation, and it is also a mapping process, a manner of arranging / fixing shapes in relation to other shapes within an area.

Every map is a fixing of some sort, perhaps even an imposition onto a place, however if sound is used to map a space there is no fixing, because sound is a temporary interaction with place, and although sound mapping is not accurate and can not reveal a truth, sound (as I have underlined previously in this paper) interacts with place revealing the spatial construct and the material condition of place.

In a work titled Foot Tunnel Driftsong (2008) I utilized an internal, outside place, the foot tunnel under the Thames River between Greenwich and the Isle of Dogs. Colleague Birgitta Hosea stood at the north end of the tunnel holding a video camera, while I began walking from the southern end sounding place.[42] The tunnel has a number of bends, dips and rises, causing sound waves to flow in a spiral around the ceramic lined tube, and to muddle together as they bounce back, forth, up and down colliding against each other. In the documentation the resonance, echo and ambient drone from voices, lift mechanics and air drafts, masks my vocal for a few minutes before it strangely rises clear, and remains at a constant volume, as though not moving at all.

Then within a few feet of the camera, the volume increases suddenly and in an exaggerated Doppler shift, just as quickly fades away. This work clearly evidences sound interaction with place, the vocal drawing is manipulated, and distorted by the space, and the vocal sound in resonance, describes the spatial dimensional, and material aspects of the cylindrical tunnel.

I asked English artist and musician Jem Finer whether he thought sound might be affected by and affect the outside environment. Finer replied "sound is affected - shaped, modulated, resonated, echoed etc - by atmospheric conditions and physical surroundings (what you call the "outside environment"). ...." and "A space, a place can feel very different according to the sound in it, threatening, benign, tranquil, welcoming . . all through sound." Finer then described Foot Tunnel Driftsong as illustrating the affect o f sound on place "perfectly actually, as peoples vocal responses to walking through it vary from the tuneful whistle through whoops and screams to silence."

[42] The Thames foot tunnel is 396 meters long, 11 meters wide, 6 meters high and 23 meters below the rivers surface. The foot tunnel was built between 1825 and 1843 by Marc Isambard Brunel and his son Isambard Kingdom Brunel http://en.wikipedia.org/wiki/Thames_Tunnel

Figure 13.10 Montreal Driftsong participant wearing the paper bag disguise. 08/09/10 still frame from moving documentation of performance m.foá

Figure 13.11 Art and Cartography workshop participants performing a bunch Driftsinging 08/09/10 still frame from moving documentation of performance m.foá

But when I asked Finer if he thought sound could map a place he told me that "sound and mapping come unstuck ...sound can't be trusted... its origin becomes muddled."

13.5 Conclusion

Although Finer believes sound cannot be trusted as a material to map place, the conversation with Georg Gartner offered the possibility that because sound does not delineate or fix place, and because place itself has a hand in the action of sound in place, sound might be the most democratic (albeit imprecise) method of mapping place.

Clearly sound is significant in the process of examining and experiencing place. It can enhance interpretation and investigations of the material and spatial conditions of place. However because cartography involves a selection of human (political and or geographical) decisions, imposed upon the arrangement of the physical world, sound does not (although technology may yet find a way to manipulate the sonic wave) conform to those decisions. Sound in relation to its source, can define areas of different materials – and also the spaces between those materials. Therefore it is plausible that a machine could be built using the voice and sonar technology to an area with

Figure 13.12 Art and Cartography workshop participants performing a line Driftsong through a Montreal car park 08/09/10 still frame from moving documentation of performance m.foá

vocal reflections and to image those vocal reflections in two dimensions (akin to sonar and thermal imagery).

In the Driftsinging process I have borrowed from the methods of Sonar, Sounding, Echolocation and the Situationist's Drift. I would like to expand the Driftsinging process by employing sonar and thermal technology, together with computer imaging software, to develop a VONAR (voice navigation and ranging) tool. The vocal sonar machine, visually evidencing sounds interaction with environment, might prompt participants to think about outside space and their relation to place differently, while also introducing the possibility (with images to visualise each vocal interaction with place), that sound might be (albeit temporarily), the most democratic method of mapping place.

Chapter 14

Sound Cartography Approaches to Urban Soundcape Research : CitySounds and Sites-of-Respite in the CBD of Melbourne.

Jordan Lacey and Dr. Lawrence Harvey

SIAL Sound Studios, School of Architecture and Design, RMIT University, Melbourne Australia

Abstract

This paper describes two related projects completed at SIAL Sound Studios at RMIT University. The first is the *CitySounds* project which was an interactive 3D Audio-Visual Cartography environment holding a community soundscape survey. The aim of the project was to investigate attitudes and awareness to a range of sounds and acoustic environments in Melbourne's CBD. The virtual environment and survey also contained propositions for future acoustic design interventions, particularly for sites-of-respite. The second project describes a recently completed pilot study, seeking suitable locations in Melbourne's CBD for actual *Sites-of-Respite*, using a combination of sound recordings, studio analysis and spatial analysis. While research approaches have been informed by listener-centered interactive representational mapping, speculations are made on future research to be informed by performative mapping, particularly emotional mapping, to augment sound design interventions to recompose the homogeneity of the *striated soundscape*.

14.1 Context Statement – *CitySounds*

Issues such as sonic identity, auditory experience and noise exposure are rarely examined in the planning and development stages of urban environments. While in part this is due to the education of designers and the low awareness of sound in the community, it is also related to available tools for representation, collaboration and communication. The two projects discussed here present approaches for mapping, representing and interacting with urban acoustic environments, ultimately for the purpose of creating urban design interventions.

CitySounds was an early project of the SIAL Sound Studios, an electroacoustic music studio based in the School of Architecture and Design at RMIT University. The Studios are a centre for cultural, design, research and educative projects into the auditory experience of the city, spatial music composition and acoustic design. The *CitySounds* project, an interactive games-engine based 3D audio-visual cartography, which was completed by the Studios in 2006, has been recognised as one of the few successful attempts at creating a graphic representation of sounds which, 'integrated sounds into a 3D city model to provide an audio-visual simulation of downtown in real terms' (Schiewe and Kornfield 2009, p.4). *CitySounds* included an embedded survey methodology, which was substantially built on the listener-centered approach of acoustic ecology, which is the study of the relationship between individuals and communities and their acoustic environment (or soundscape). The purpose of the survey was to provide Melbourne City Council with a broad scope of information to help with the development and implementation of noise management initiatives. While the City already had acoustic measurement data and noise complaint data as indicators of sonic qualities, the survey aimed to provide Council with detailed information on community attitudes and awareness to a range of sounds and sonic experiences in Melbourne's CBD.

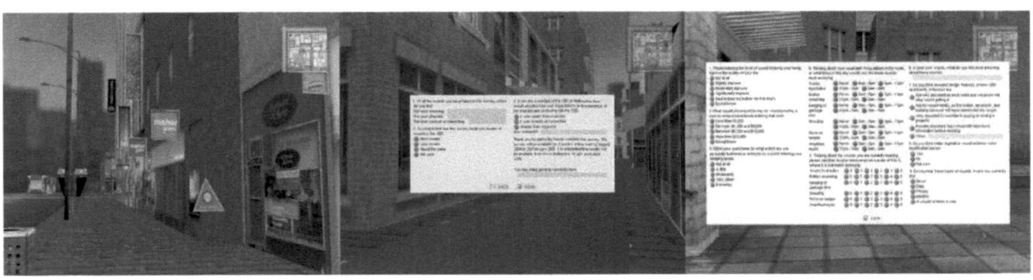

Figure 14.1 Images of *CitySounds* Interactive Map

Three images from *CitySounds*, an interactive games-engine based 3D audio-visual cartography experience created by SIAL Sound Studios. Note survey questions.

The approach combined three main elements inside of a games engine: a 3D model of an indicative CBD precinct, a soundscape and an embedded community soundscape survey. A report was generated from the survey results[1]. Table 14.1 shows the main stages of the *CitySounds* project. The table is not strictly ordered in time, as stages overlapped or were undertaken in parallel. The aim of mapping urban sounds into the games engine was to maintain contextual elements (sounds, typologies, visual features) of actual environments while simulating the episodic structure and accumulation of aural experience of a soundscape.

Table 14.1. CitySounds project stages

Stage	Process or activity
1	Consultation, assessment and determination of site types.
	Consultation with City of Melbourne staff to establish criteria for virtual precinct creation and soundscape design.
	Listing of actual sites for modeling and audio recording.
	Initial assessment of sites at different times of day by project team.
	Initial sound recording and digital image capture as the basis for indicative site modeling.
2	Sound, image and data capture.
	Digital audio recording.
	Digital image capture.
	Texture capture and library creation.
3	Virtual environment construction.
	Production of 3D model of site, including selected interiors.
	Soundscape production in parallel with questionnaire development.
	Optimisation for on-line delivery and testing.
4	Research questionnaire.
	Research previous local and international community noise surveys.
	Development of community response questionnaire.
	Development of on-line delivery system for results.
5	Survey Active.
	Live release and maintenance for seven months.
6	Survey Report

This table shows the main stages of the *CitySounds* project. The table is not strictly ordered in time, as stages overlapped or were undertaken in parallel.

The final *CitySounds* model and survey was launched on the 12th of August, 2004 and made available to the public approximately one week later. The 3D model was released online, made available on CDs, and could be accessed at city-based libraries. The survey was 'live' for seven months, closing on the 31st of March 2005.

[1] The report is available at http://www.rmit.edu.au/browse;ID=nebs78ggixks. An extensive discussion is available in Chapter five of Harvey's PhD thesis, available at http://adt.lib.rmit.edu.au/adt/public/adt-VIT20091029.121056/. For a discussion of games technology and acoustic ecology, see Harvey and Moloney, 2004.

Analysis and writing of the Report were completed in the second half of 2005, and the Report officially released in early 2006. Approximately six hundred people responded to the survey generating 3,949 reportable results. The final project was approximately one and one half city blocks in size, and contained around fifty sound files. The thirteen-member project team included sound designers, an acoustician a social scientist, computer programmers and 3D model designers.

The model was designed to provide users with an indicative[2] real-time sonic landscape (a soundscape) containing sonic environments they might encounter in their daily traversing of a city precinct. The issue with most community sound surveys is that participants are expected to conjure from their aural memories sonic events such as aircraft landing, traffic flows or construction noise and then respond to questions about their experience of those sounds at particular times. The 3D environment and soundscapes sought to provide for participants experiential cues or markers for discrete locations, not accurate simulations, but enough sonic material to avoid long text descriptions to 'set the scene' for survey questions. Survey participants visited nine sites, or general space types, and their associated soundscapes while completing the questions in the model. Users could select from two navigation modes: 'guided-tour' which automatically moved to all sites or they could self-navigate to some or all sites, meaning participants had the opportunity to select between representational and performative mapping experiences[3]. Individual's survey results were sent to servers at RMIT, and later collated by researchers for analysis and reporting.

14.2 From *Citysounds* to the Sites-of-Respite Pilot Study

The influence of *CitySounds* emerged in the Melbourne City Council *Towards a better 'Public Melbourne', Draft Urban Design Strategy.* Drawing on research from *CitySounds* the report calls for strategies and opportunities for the design of sites-of-respite, which would afford people more prospects to pause in the city, including a recommendation for a:

> ... *'schedule' of areas that could be developed in the longer term as quieter pedestrian thoughfares and quiet spaces for the general public* (City of Melbourne 2006, p. 37).

This recommendation was developed in relation to the survey finding that 70%

[2] The model was indicative in that sounds were not presented at the same loudness or precise spatial relationships as would be encountered in real space.

[3] This is not strictly true of course as performative mapping suggests that '(n)ew worlds are made every time a map is deployed'(See Perkins 2009, p.1); however, the experience did accommodate for some agency through choice.

of respondents indicated they would use sites-of-respite at least once a week if such sites were available, which subsequently led to the development of the pilot project, *Sites-of-Respite* research presented here. The *CitySounds* survey also contained questions to investigate other significant information to determine where sites-of-respite might be located. This included questions to determine how long people would walk to access a site, times of day a site should be accessible, types of activities a site might afford e.g. relaxation, conversation, reading or reflection. Using the survey results and the authors' acoustic knowledge, we designed the pilot study to investigate the potential of Melbourne's network of laneways as locations for one or more sites-of-respite.

Author Lacey primarily conducted the study during an internship from July to November 2010. Some site visits and selection occurred prior to this period however the main components of the research were conducted in late 2010. The research detailed below involved ten primary research activities encompassing a literature search, citywide site visits and a process to rate sites via an attribute list, field recordings, generation of sonograms and analysis of results. Speculations for future work appear at the end of this chapter, while other work planned includes further cartographic investigation and establishment of prototype sites.

14.3 Literature Search

A literature search was conducted alongside the field and studio work. The purpose of this was to identify gaps, trends and nodes of activity within contemporary soundscape studies. The literature search sought previous work on sound in urban environments and focused on publications across urban design, public health, acoustics, soundscape studies and cartography. For the purposes of this paper only two aspects of the literature search will be discussed: cartography and sound-based design. Other interesting texts are referred to in the speculations section at the end of the paper.

Caquard et al. (2008) note that 'the combination between maps, sound and technology remains unusual' (p.1228); however, there have been some key papers arguing for the development of a sound cartography. Smith (1994) in her paper *Soundscapes* argued that, 'social geography could be enlarged and enriched if sound were more explicitly incorporated into their remit' (p.233). She only refers to the World Soundscape Project (WSP) briefly, describing the project as archivists rather than ethnographic[4]. Smith approaches soundscape mapping as an exercise of

[4] This is not an accurate portrayal of the WSP research. While recording and archiving where necessary to capture the material of the soundscape for further study, the WSP were also highly active in educating the community through multiple listener centered activities, including sound-

understanding culture through musical diversity. Smiths approach to soundscapes is not commensurate with the approach of acoustic ecologists, who consider music a *component* of the soundscape, not *the* soundscape. The soundscape approach as conceived by the World Soundscape Project is an attempt to develop,

> *...acoustic design (which) is to regard the soundscape of the world as a huge musical composition unfolding around us ceaselessly... (and that) acoustic design... is a matter of the retrieval of a significant aural culture and that is a task for everyone* (pp. 205-6).

Another often cited work is *Sound and Geographic Visualisation* by J.B. Krygier (1994), which deconstructs sound into, 'a basic set of abstract sound variables... which are viable for geographic visualization applications' (p.4). This approach is untenable to a soundscape approach as the soundscape is considered as a whole. Krygier's intention is not to understand the soundscape, but to use sound for the purpose of representing variables (sound or otherwise). Sound variables within soundscape studies can be considered as sound sources and listener relationships with those sound sources, which may have value in representational mapping. For example, the Positive Soundscape Project (2009) collected descriptive variables of people's emotional relationship with sound by collecting data from their experiences of moving through simulated and/or real soundscapes:

> *'(a) question asked of the listeners... was "How does the soundscape make you feel?" and the rating scales used were: calmness & relaxation, comfort & reassurance, vibrancy & arousal, informative, intrusiveness, pleasantness* (p.3).'

A more contemporary text by Caquard et al. (2008) argues for an increasing use of sound in cartography including voice, music, sound effects and soundscapes. The authors go on to say that,

> *'...sound design... should not be viewed as simply a convenient way of conveying more descriptive and quantitative information about space and territories. Rather, the use of sound forces us to rethink the very concept of the map as primarily a visual image of space that serves as a simple conveyer of information. The integration of sound might involve a deeper understanding of the cultural, geographical and political dimensions of maps'* (p.1228)

This sentiment could be read as being disagreeable to the methodology of *CitySounds*, the primary intent of which was to extract quantitative survey data for policy and planning by government; however, parallel to this was the potential for listeners to immerse themselves in a simulation of the Melbourne CBD and help inform policy that could potentially transform their 'real-world' listening experience. The authors therefore disagree with the sentiment that sound cartography should not be about presenting and collecting quantitative information as this is

walks, and conducted several transcultural projects involving mapping and data collection. See the Appendices in Schaffer (1977) and Torigoe's (1982) thesis.

essential to transforming the cultural, geographic and political dimensions of a city, and thence the communities potential future performative mapping experiences. In the hands of a skilled sound designer soundscape-cartography can convey both quantitative and qualitative information.

Papers describing the difference between representational and performative mapping were investigated. Of particular interest were Perkins (2009) and Turnbull (2007), who discuss the shift from traditional representational mapping to performative mapping, which can take on a variety of forms but centralises the importance of the human body and human experience in the emergence of maps. These texts are elaborated on later in the paper.

An important theme to emerge for the authors from the literature search were methods in recomposing the soundscape with natural (e.g. water, birdsong) (Jeon 2010) and artificial (electroacoustic) means (Hellstrom 2008) to both enhance the soundscape and to mask unwanted sound. These texts refer to the critical work of sound designers, whose transformations can diversify the urban soundscape, creating opportunities for listeners to enjoy unfolding exploratory experiences and an increasingly totalised engagement of the senses. This is discussed at length below. Biophilia and biohpilic design were investigated at length, and will inform future research. This is a controversial hypothesis and as such its analysis is beyond the scope of this paper. For further reading on biophilic design see Kellert et al. (2008).

14.4 Project Work

The *Sites-of-Respite* pilot project was conducted over approximately a four-month period. Table 14.2 outlines the different stages of the project. Following is a detailed explanation of the research undertaken.

Table 14.2. *Sites-of-Respite* project stages

Stage	Process or Activity
1	Initial exploration of the CBD
2	Compiling site descriptions and photos into initial report
3	Initial choice of 35 sites
4	Creation of Attribute list
5	Bounding exercise to select six sites using attribute lists and further site visits

This table shows the stages of the *Sites-of-Respite* research. Stages are in sequential order except for the literature search, which was ongoing throughout the project.

Stage	Process or Activity
6	Three recordings of each site – 20 min outside, transition recording, 10 min inside (3 hrs total recording time)
7	Studio listening and analysis
8	Sonogram generation and analysis
9	Literature search and its contextualization within field and studio work
10	Generation of report

This table shows the stages of the *Sites-of-Respite* research. Stages are in sequential order except for the literature search, which was ongoing throughout the project.

14.4.1 Identification of Sites

In the initial stage of this project Melbourne's CBD laneways were explored on foot. The boundaries encompassing the explorations were Spencer St, Latrobe St, Spring St and Flinders St. Every laneway in Melbourne was visited. Arcades, squares, and other sites such as alcoves and underpasses were also visited. We were particularly interested in sites that were protected from major arterial road noise and therefore had greater potential to be established as areas of respite from city noise. In this initial stage potential site locations were simply marked on a map of the CBD taken from the local street directory, The Melways.

14.4.2 Initial visiting of the Sites

Exploring the lanes took approximately one week. In this time photographs of most of the sites and notes on our immediate impressions of the sites were collated. If sites were obviously unsuitable they were not recorded. Sites were considered unsuitable if they had no protection from traffic noise, if they had exceedingly loud air-conditioning noise, or if the site was crowded with rubbish bins, or with people sitting in street cafes. Arcades were immediately excluded from consideration. This is due to the arcades being crowded with people, and exclusively set aside for commercial purposes. Due to the hard surfaces and closed roof and excessive noise sources, these sites tend to be noisier than outdoor laneways; although they do have their own acoustic qualities, they are unlikely for use as sites-of-respite. Parks and gardens in the CBD were also excluded; unlike laneways they have no protection from external noise – while they are visually appealing, they are not suitable as sites-of-respite[5]. This reduced the potential number of sites to 35 squares and lane-

[5] There are some interesting artificial soundscape installations designed to mask noise issues and

ways bounded by substantial built forms.

14.4.3 Creation of attribute list and final selection of sites

As the project was a pilot study, a method was needed to de-select some of the 35 potential sites so that available time and resources could be used for a comparative assessment of indicative sites. The selection process was undertaken through subjective analysis via field notes and a photo gallery. We then developed and applied an initial attributes list to set bounding conditions that would allow us to identify the best sites for further audio recording and computer-based analysis. This list was implemented to try and obtain a clear idea of which sites were going to be most suitable as sites-of-respite. This was to acknowledge that it was not just the acoustic qualities that would make a site suitable. Other attributes such as smells, visual affect, ambience, access to traffic, natural features, dampness, use by people, and suitability for electroacoustic performance were also considered.

Each attribute was marked according to a scale. Analysing these scales led to the selection of the final six sites: Spring Street, Baptist Place, Drewery Place, Little Bourke Place, Manton Lane and McLean Alley. It is important to note here that the attributes list is an extension of the initial mapping exercise. Using location-based media, we can imagine each attribute list attached as a sort of meta-data list from the site location of the map, giving us a deeper understanding of site conditions. Virtual mapping would allow deeper levels of interpretation through computer navigation, and there is potential for the development of mobile phone applications for on-site interpretation[6].

14.4.4 Field Recordings

The next step was to take field recordings of the six selected sites. We used a Sounddevices 2 channel 722 digital audio recorder which gave approximately 45 min recording in the field, and two Rode NT5 microphones, which were mounted on a stand to achieve stereo imaging. Three recordings were completed at each site:

- 10 minute recording outside of the proposed site
- transition recording moving into the site
- 20 minute recording inside the site.

We determined that a 10-minute recording would be the minimum. This length

---

enhance listener experience in parks and gardens. For example see Hellstrom 2008.

[6] An interesting development along those lines is broadcastr, an online map based platform for pinning oral histories to google type maps. See < http://www.broadcastr.com/>. Accessed 2 February 2011.

ATTRIBUTES FOR SELECTING SITES OF RESPITE		
LANEWAY:		
VARIABLE	**SCALE**	**COMMENTS**
Rubbish Bins (Small & Large)	Yes No	
Ground Litter	Yes No	
Traffic access	Yes No	
Parking allowed/Loading zones	Yes No	
Car park entry	Yes No	
Excessive noise - air-con	L---------------H	
Excessive noise - traffic	L---------------H	
Geographic location relative to users	Near-----------Far	
Business located inside	Yes No	
Residental homes adjacent	Yes No	
Natural Features	L---------------H	
Graffiti	L---------------H	
Use by Tourists	L---------------H	
Use by Recreational users	L---------------H	
Dampness	L---------------H	
Smells	Weak-------Strong	
Ambience (atmosphere)	Still---------Active	
Possiblity of Seating areas	Yes No	
Suitable for electroacoustic performance	Yes No	
OTHER COMMENTS		

Figure 14.2 Attributes List

The attribute list was used to set bounding conditions to reduce in a meaningful way the number of sites subjected to analysis to six sites.

was chosen to allow a suitable sized sample for later analysis in the studio, which would allow us to determine some of the patterns or repetition of sounds occurring in the soundscape. Twenty minutes was chosen for recording inside the site, as it approximated the duration a listener might visit a site-of-respite: shorter than a

Figure 14.3 Six Sites Selected for Field Recording and Studio Analysis.
The six sites selected in Melbourne's CBD for field recording and analysis were Spring Street, Manton Lane, Drewery Lane, Little Bourke Place, McClean Alley and Baptist Lane. All photos were taken inside the sites except for the Baptist Lane photo, which was taken outside of the site.

lunch-hour but longer than a 'drop-in' visit. The longer time period allows greater opportunity to recognise changes in the soundscape, and provides more data to analyse in the studio. Transition recordings, which involved carrying a microphone through a streetscape, were difficult at times. Wearing squeaky shoes or a rain jacket that rustled, money jingling in pockets, passers-by yelling into the microphone and gusts of wind caused problems in studio analysis and meant certain sites had to be re-recorded. Some transition recordings had to be completed several times to avoid these problems. At certain times rain or excessive wind caused field recordings to be cancelled until more suitable weather returned. For these reasons, the six site recordings took five weeks to complete.

14.4.5 Recording Analysis

Once all the recordings had been collected they were downloaded in the Studios and recreated for sonogram analysis using the program, Audiosculpt[7]. A sonogram analysis (frequency spectrum change over time) of the recordings, with a frequency range of 0-20kHz, was completed. Alternatively, transition recordings were analysed in a spectrum VST in Ableton Live[8]. All transition recordings had the same characteris-

[7] For an introduction to the Audiosculpt program see < http://forumnet.ircam.fr/691.html?L=1>. Accessed 2 February 2011.

[8] For an introduction to the Abelton Live program see < http://www.ableton.com/>. Accessed 2

tics in that predictably they showed a marked decrease in middle and high frequencies through the transition, as the brick walls of the laneways were attenuating these frequency ranges; however, the lower frequencies (100Hz and below) had little to no reduction. This can be accounted for by the reduced attenuation rates of lower frequencies by the brick walls of the laneways, and also the high number of air-conditioners in the laneways with low-frequency output. Following is an example of a list of findings describing the recordings inside and outside of the six sites.

Table 14.3. Studio notes for Drewery Place recording

Drewery Place			
Inside	Outside	Compare In&Out	Transition
Air-conditioner dominates space with bands at 75/150/300/450 Hz with quieter 150 Hz overtones up to 2000 Hz Abundant and active bird life with bird song ranging from 3000 – 8000 Hz particularly at 6-8 min Few car sounds but could be time of day (4pm) Door slam at 16m30s fills the whole spectra. This is from a restaurant storeroom that opens onto the site. Any opportunity for reflection is drowned out by the constant air-conditioning noise	Pedestrian traffic-crossing signals at 2700 and 8200 Hz periodically Generally dense noise at 0 – 2000 Hz Air-con noise at 300 Hz still prevalent but masked by sounds of traffic/crossing lights	A huge difference. Inside a definite respite from the sounds of traffic and crossing signals; however the air-conditioning noise reduces respite advantages	Reduction in mid and high frequencies as transition into site occurs. Low end frequencies still strong throughout the recording

Detailed notes of information collected during studio analysis of field recording of Drewery Place, using Audiosculpt.

February 2011.

14.5 Observations from Analysis

Through immersion in the six recording locations and extensive sonogram analysis of the eighteen recordings some consistent patterns could be deciphered. These patterns were recorded as observations and are distilled into seven broad themes, which are listed below. These observations will inform future research. The relevance of observations to cartography are placed in descending order, with most emphasis placed on the last point, *the Striated Soundscape*, which will inform future project work at the Studios and is pertinent to this papers description of the relationship between sound design and cartography.

1. *Unbounded space: 1 Spring St*

With the exception of 1 Spring St the traffic noise always decreases dramatically in Melbourne's laneways. This is unsurprising, as 1 Spring St is not surrounded with high walls that could attenuate traffic noise.

2. *Auditory niche of birdsong*

Bird life appears to react to high-pitched frequencies as seen in 1 Spring St and McClean Alley, such that the birds alter the pitch of their song to account for incoming frequencies that lie within the frequency range of their birdsong. Figure 14.4 provides an example of this phenomena, where birdsong occupies the 3-4khz frequency band, therefore avoiding the repeating mechanical sound at 4khz. In most other recordings birdsong inhabited a niche of 3-8 khz, as demonstrated in the studio notes for Drewery Place as shown in Table 14.2. This phenomena has been written about extensively in *Urban bioacoustics* by Warren, Katti, Ermann and Brazel (2006). For a recent study in Australia see Potvin (2011).

3. *Natural sounds in the CBD*

Natural sounds in Melbourne's laneways are - birdsong, wind in leaves, human voices. (Possum calls and bat calls are audible at night in Melbourne's parks and gardens, but this was not relevant to the sites we investigated).

4. *Comparison of natural sound and mechanical sounds in the CBD*

The complexity of the patterns of the sonogram of birdsong and human voices (natural sounds) contrast strikingly with the flatline[9] sounds of machinery and the noise of traffic. These are striking visual representations of the diverse aural experience of natural sounds compared to the constant and repetitive sounds of machinery. The above three observations are most important in the context of biophilic sound design research.

5. *Low frequency hum*

Most notable in all the recordings is the ubiquity of low-frequency hum

[9] Flatline or Stationary sound, is one that is relatively unchanging. See entry for Stationary Sound, at <www.sfu.ca/sonic-studio/handbook/index.html>. Accessed 6/5/2011.

throughout the city, particularly from traffic noise and air-conditioning noise. Many papers have been written on the connection between low-frequency noise and health. An extensive bibliography appears in Goines (2007).

6. *Laneways and Respite*

Laneways provide respite from most city noises except for low-frequency sounds. Other exceptions are situations where air-conditioners and ventilation systems are located in or very close to the selected sites.

7. *The Striated Soundscape*

A careful analysis of the bottom portion of Figure 14.4 will reveal a series of straight lines running through the sonogram which are in fact the sounds of the all-pervasive sounds of air-conditioners and cooling plant systems for large computers operating in city buildings. Also the diagonal and vertical straight lines represent sirens and the roar of cars. Deleuze and Guattari[10] in their book Nomadology (1986), use the concept of striated space as a metaphor to describe the domination of space, particularly open space inhabited by nomadic cultures. The striations referred to in *Nomadology* refer to the built environment such as the aqueducts and pillars of ancient Rome or the highways and grid-lines of the modern city. Deleuze and Guattari write of striated space:

> *'Homogeneous space is in no way a smooth space; on the contrary, it is the form of striated space. The space of pillars... These parallel verticals have formed an independent dimension capable of spreading everywhere, of formalising all the other dimensions, of striating all of space in all of its directions, so as to render it homogenous'* (p.32).

Debord (1955), the originator of the term pyschogeography, also suggests such a domination of space when he criticizes,

> *'Haussmann's urban renewal of Paris under the Second Empire (which was) motivated by the desire to open up* broad thoroughfares *allowing for the rapid circulation of troops...'* (p.1). [Author's emphasis]

Additionally to these descriptions of the domination of space by striations, Henri Lefebvre (1991) in the *Production of Space* also refers to the domination of space by Modernity. He writes:

> *'A society is a space... whose abstract truth is* imposed *on the reality of the senses, of bodies, of wishes and desires'* (p.139).' [Author's emphasis]

[10] The reference to Deleuze and Guattari may come as a surprise to the reader. Deleuze and Guattari usually emerge in a post-structuralist context and certainly not in the realm of data collection as has informed much of the work described in this paper. However, critical theory is best understood as a tool box from which concepts can be appropriated and developed. As explained by Deleuze: 'A theory is exactly like a box of tools... It must be useful. It must function. And not for itself. If no one uses it, beginning with the theoretician himself... then the theory is worthless...' (Foucault, 1977, p. 208).

So what is the connection between these (radical) philosophical observations and the observations emergent in the sonograms? As mentioned above, an analysis of the lower section of the sonogram in Figure 14.4 reveals the homogenising striations of the modern urban soundscape. It is likely that recordings of many modern cities would reveal a similar pattern if only for the reason that the mechanisations that create such sound are global (air-conditioners, plant rooms, cars, emergency vehicles). The outcome of this is that listeners are deprived of a sense of place, as the homogenizing mechanical agents of modernity dominate space. A sense of place is achieved when, as explained by Truax,

'*...the flow of communication goes both ways, since the listener is also a sound-maker, and therefore it is the entire system of the listener plus environment that constitutes the soundscape*'(p.65).

Therefore a sense of place, aurally, requires the ability of listeners to project themselves into space, impossible in the modern soundscape which is loud, homogeneous and 'fixed'[11]. A sense of place also requires a diverse acoustic ecology, which in turn creates a diversity of environments from which a sense of place can be felt, and from which a sense of place can emerge. Stephen Feld in his work with the Kaluli people of the Bosavi in Papua New Guinea describes sense of place beautifully:

'*...places make sense in good part because of how they are made sensual and how they are sensually voiced. Poetic and performative practices centralise the place of sense in making a local sense of place*'(p.133).

This again emphasises the challenge of developing a sense of place in the modern city where, if the sonogram is to be regarded as metonymy for the modern soundscape, the homogenising effects of the striated soundscape render each city to nearly the same sonic conditions and thus the same listening experience. Perhaps Smith's emphasis on mapping the music of urban environments rather than the soundscape of those environments makes more sense when looked upon in this context. To refer to another idea from Deleuze's tool box, the emerging research approach to design in this project is to create a soundscape that is rhizomatic, and not arborescent[12], so that the listener has a way *in* to the experience of listening, of belonging, and of creating. As discussed earlier, developing tools for removing homogeneity and promoting diversity is the work of the sound designer. In the following section we hope to take the first steps in developing some theoretical tools that merge design and cartography, in particular performative cartography (Perkins 2009, p.5)[13].

---

[11] Atkinson (2007) refers to the 'stickiness of sound' when describing the acoustic ecology of a city. He uses this observation in his argument for the creation of an aural-geogrpahy (p.1905).

[12] Perkins (2009) desribes rhizomatic mapping as a 'powerful, open-ended, indeterminate process that unfolds and makes new places and actions' (p.2). Aborescence in the modern soundscape could be used to describe the *affect* of striations – a totalising principle.

[13] It should be noted that there might be some protest from the reader that the lines representing

14.6 Speculations on Future Work

The relationship between cartography and sound has traditionally been dominated by acousticians and SPL measurements, giving us predictable maps of red and blue related to high and low decibel measurements, a clear example of representational mapping. This relegates soundscapes to a condition of noise-is-volume as measured against a scale from which government agencies can formulate and enforce policy, and ignores the central importance of people's relationship with the soundscape. Such negative connotations associated with the soundscape are being challenged as is evidenced by the Positive Soundscape Project in the United Kingdom, which seeks to centre soundscape analysis on the impressions of urban dwellers (Cox 2010). *CitySounds*, as described above, is another example of placing listeners at the centre of soundscape analysis.

Developing performative modes of mapping could be highly beneficial to the work of the acoustic ecologist and sound designer. The acoustic ecologist listens and encourages others to listen. In this mode we are asked to recompose our soundscape so it reflects us, as individual and community, and not dominated space, as reflected in the *striated soundscape*. Performative mapping suggests a way that the listener can re-compose their soundscape as they move through the city so that it reflects their sense of self and place. This is becoming increasingly difficult in the homogenised urban soundscape; however, it has been achieved to some extent with the iPhone application RJDJ[14], which empowers the user to transform their soundscape as they move through it, and the Sonic City project which turned the city into a musical instrument in a fascinating (perhaps unintended!) example of performative sound mapping (Gaze et al. 2003).

Important to the development of a link between soundscape design and performative cartography is the notion of affordances. Affordance is an important environmental psychology concept that breaks down the distinction between subject and object and thus has relevance as a design tool when considering relationships between soundscape design and performative cartography. The philosopher, Gernot Bohme, describes the atmosphere between the listener and a sound source as a corporeal substance, where 'atmospheres constitute the "In-between" between environmental qualities and human sensibilities' (2000, p.15). He goes on to say that '(a)tmospheres stand between subjects and objects: one can describe them as object-like emotions, which are randomly cast into a space' (p.15). A link is evident between Bohme's description of Atmospheres and the Theory of Affordances as described by Gibson (1986), which 'implies the complementarity of the animal

bird song in the sonogram appear striated. In fact they are not. With the convenience of a zooming function it can be seen that these are broken lines; additionally, the dynamics and tempi show variability.

[14] See http://rjdj.me/

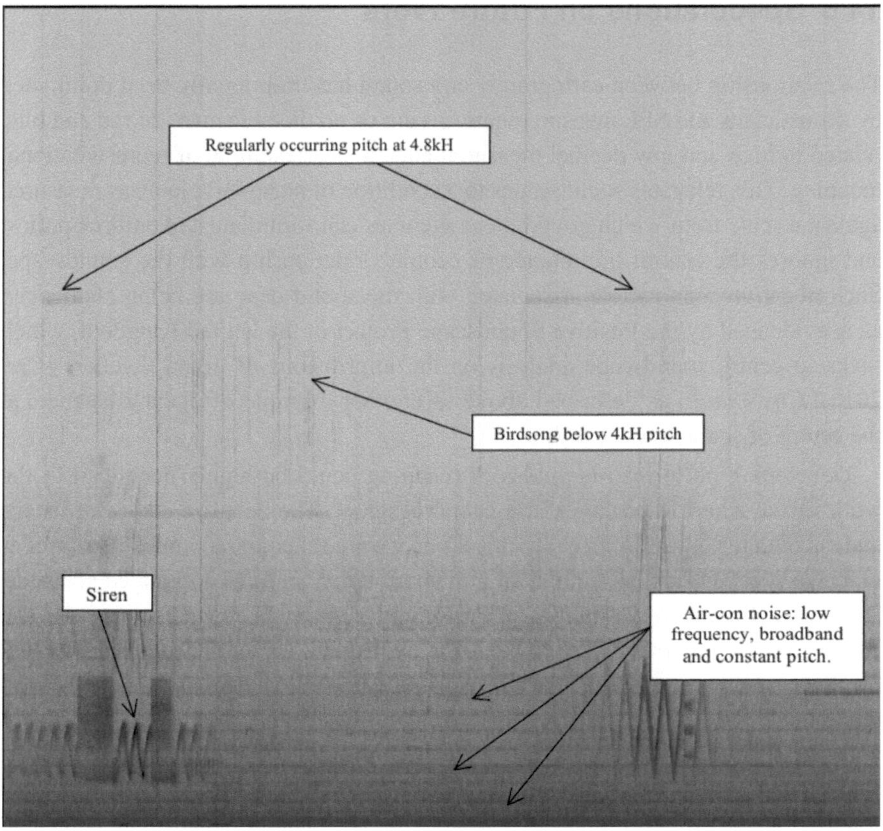

Figure 14.4 Sonogram of McLean Alley

This is a section of the sonogram generated from the recording inside Mclean Alley. It shows a niche in the soundscape around 3-4khz, which is occupied by birdsong. Note striations in the bottom section of the figure. These have been labeled as sirens and air-conditioning noise.

(including humans) and the environment' (p.127). Thus the soundscape and the listener merge in the "in-between", where simultaneously an affordance appears; whether the soundscape takes on a dull and oppressive atmosphere, a vibrant and calming atmosphere, or otherwise is interdependent with the listener's emotional experience of sounds. Emotional cartography approaches could be developed to encourage members of a community to create maps of their own emotional experience when encountering sound (as described above the Positive Soundscape Project has developed this idea, though as a data collection activity rather than a mapping technique).

It is plausible that such performative mapping could inform representational mapping created as tools for diversity, vis-a-vis fragmenting the striated soundscape; remembering diversity and the relationship between the diverse is the essential ingredient of a healthy ecology. Cartographic representations could map a city soundscape, reflecting both its content and spatial organisation, which when overlaid with listeners' emotional maps could provide rich informational records of human experience in urban soundscapes. Such materials used as research for design, could assist acoustic designers, urban planners and composers to diversify recognised homogeneity in the soundscape, thus affording new acoustic ecological relationships between listeners and their city. Examples of this can be found in the architecture of play[15] in which unexpected art installations can enrich and diversify the experience of people moving through the city, such as Skyhooks where invisible sounds can be triggered by people's movement (Franinovic & Visell, 2007, p.194).

This research represents the first stages in a process where compositional notation and mapping might enter into a new relationship; a new discipline, predicting and representing uniquely designed micro-acoustic environments, periods of quiet, of silence, transitions and diverse environments, where the movement across acoustic boundaries alerts the mind and stimulates the senses. If, as sound designers, we can encourage heterogeneous sound environments we may open the way for the performative mapping potentials of rhizomatic and hodological[16] experience. Melbourne's CBD is particularly well configured for these kinds of experiences due to its unique network of paths and laneways (COM 2006, p.11). If we move through the city and hear only one type of sonic quality (as is the case with the striated soundscape), the same irritation and boredom may be the consequence as if one is forced to sit through an unchanging stream of white noise. Varying acoustic environments could potentially diversify the soundscape experience just as a piece of music that shifts in timbre, pitch and dynamics is of more interest to the listener. In this sense sound cartographies are spatialised soundscapes viewed as a "written" composition, where composition is an affordance inseparable from listener experience and interactivity; as the individual moves through the soundscape, composition comes to life as her ears receive sound and she returns sound. This is in contrast to the striated soundscape in which the listening experience does not contain affordances as it does not consider the listener. As explained by Truax:

> *'(There is)...an increasing standardisation and commoditisation of the listening experience. The mass product may be a technological extension, but it sounds the same everywhere in the world, every time it is reproduced. Its consumption simplifies the relationship of the listener to sound and contributes to a homogeneity of the soundscape.'* (Traux 2001, p.163)

In conclusion it must be reiterated that this paper describes an unfolding process

[15] For an extensive explanation of architecture as play see Franinovic (2010).

[16] Hodology refers to the creation of paths and trails through human agency. In hodological mapping knowledge and space are co-produced, such that we 'make our world in the process of moving through and knowing it' (Turnbull 2007, p.142).

at SIAL Sound Studios, which includes an emergent interdisciplinary synthesis between sound design and cartography. Unlike previous works on sound cartography (as described in the literature review) we do not wish to use sound as an indicator of variables, or to analyse music as a cultural phenomena. We intend to use sound more in the context of Caquard et al. (2008) who developed an aural-cartography to inform education practice; an approach reflected to some degree in *CitySounds*, although the ambitions of these projects differ. For the mapping experience to belong to the body again, as experienced through a sense of place, then homogenising agents that spread the *striated soundscape* globally must be directly challenged through creative soundscape design, as informed by spatial analysis as described in the *sites-of-respite* project. This analysis can extend into amalgamation with performative, particularly emotional, mapping approaches, that identify expressions of homogeneity, such as the banal, boredom or the monotonous, and begin to make emergent a soundscape that is reflective of the agency of those who inhabit the soundscape, not just dominated space as expressed by the contemporary urban soundscape.

References

Atkinson, R. 2007. "Ecology of Sound: The Sonic Order of Urban Space." *Urban Studies* 44(10), 1905-1917.

Benterrak, K., Muecke, S., and Roe, P. 1996. *Reading the Country*, Fremantle Arts Centre Press, Australia.

Bohme, G. 2000. "Acoustic Atmospheres: A Contribution to the Study of Ecological Aesthetics." *Soundscape*, Vol 1, Number 1, Spring 2000, pp. 14-18.

Caquard, S., Brauen, G., Wright, B. and Jasen, P. 2008. "Designing sound in cybercartography: from structured cinematic narratives to unpredictable sound/image interaction." *International Journal of Geographical Information Science* 22:11, 1219-1245.

Cain, R., Carlyle, A., Davies, W.J., Hall, D.A., Hume, K.I. and Plack, C.J. 2009. "The postive soundscape project: A synthesis of results from many disciplines." *Internoise 2009*, Ottawa, Canada.

Cox, T. 2010. Beyond decibles: Planning the new sounds of the city. *New Scientist*, issue 2771, 02/08/2010.

City of Melbourne. 2006. *Towards a Better 'Public Melbourne': Draft Urban Design Strategy July 2006.* http://www.melbourne.vic.gov.au/AboutCouncil/PlansandPublications/strategies/Pages/Urbandesign.aspx. Accessed: 6/5/2011.

Deleuze, G. and Guattari, F. 1986. *Nomadology*, Semiotexte, France.

Debord, G. 1955. "Introduction to a Critique of Urban Geography." *Les Levres #6*. <http://library.nothingness.org/articles/SI/en/display/2> Accessed: 2/5/2011.

Feld, S. 1996. "Waterfall of Song: An Acoustemology of Place Resounding in Bosavi, Papua New Guinea," in *Senses of Place*, School of American Research Press, New Mexico.

Foucault, M. 1977. "Intellectuals & Power: A conversation between Michel Foucault and Gilles Deleuze," in *Language, Counter-Memory, Practice: selected essays and interviews by Michel Foucault* Bouchard, D.F. (ed.), Cornell University, New York.

Franinovic, K. 2010. "Architecting play." *AI & Society*, 26:2, 129-136.

Franinovic, K. and Visell, Y. 2007. "New Musical Interfaces in Context: Sonic Interaction Design in the Urban Setting." *Proceedings of the 2007 Conference on New Interfaces for Musical Expression*, New York, USA.

Gaze, L., Maze R., and Holmquist, L.E. 2003. "Sonic City: The Urban environment as a Musical Interface." *Proceedings on New Interfaces for Musical Expression*, Montreal, Canada.

Gibson, J. 1986. *The Ecological Approach to Visual Perception.* Lawrence Erlbaum Associates, London.

Goines, L. and Hagler, L. 2007. "Noise Pollution: A Modern Plauge." *Southern Medical Journal*, Vol. 100, pp. 287-294.

Harvey, L. and Moloney, J. 2004. "Visualization and 'Auralization' of Architectural Design in a Game Engine Based Collaborative Virtual Environment." *Proceedings of the Information Visualisation, Eighth International Conference on Information Visualization.* Available from <*http://ieeexplore.ieee.org/xpl/freeabs_all.jsp?arnumber=1320236*> [Accessed 02/05/2011].

Hellstrom, B, Nilsson, M, Becker, P, and Lunden, P. 2008. "Acoustic Design Artifacts and Methods for Urban Soundscapes." *15th International Congress on Sound and Vibration*, Korea.

Jeon, J.Y. 2010. "Perceptual assessment of quality of urban soundscapes with combined noise sources and water sounds." *J. Accoust. Soc. Am.*, Vol 127, Issue 3, pp. 1357-1366.

Kellert, S.R., Heerwagen, J. and Mador, M. 2008. *Biophilic Design*, John Wiley & Sons, Canada.

Krygier, J.B. 1994. "Sound and Geographic Visualisation," in MacEachren A. and Taylor D.R.F. (eds.). *Visualisation in Modern Cartography*, pp.149-166. Pergamon, New York.

Lefebvre, H. 1991. *The Production of Space*. Blackwell Publishing Limited, USA.

Perkins, C. 2009 "Performative and embodied mapping." In *International Encyclopaedia of Human Geography*, Kitchin, R and Thrift, N. (eds.), Elsevier, London.

Potvin, D.A., Parris, K.M. and Mulder, R.A. 2011. "Geographically pervasive effects of urban noise on frequency and syllable rate of songs and calls in silvereyes (Zosterops lateralis)." *Proceedings of the Royal Society*, <http://rspb.royalsocietypublishing.org/content/early/2011/01/05/rspb.2010.2296.abstract> [Accessed 19/01/2011].

Schafer, M. 1977. *The Soundscape and the Tuning of the World*, Destiny Books, Canada.

Schiewe, J. and Kornfield, A. 2009. "Framework and Potential Implementations of Urban Sound Cartography." *12th AGILE International Conference on Geographic Information Science 2009*, Leibniz Universitat, Hannover, Germany.

Smith, S.J. 1994. Soundscape. *Area*. 26:3, 232-240.

Spatial Information Architecture Laboratory. 2006. CitySounds. <http://www.sial.rmit.edu.au/Projects/City_Sounds.php> [Accessed: 02/05/2011].

Torigoe, K. 1982. *A Study of the World Soundscape Project*. MastersThesis, Graduate Studies of York University, Canada.

Truax, B. 2001. *Acoustic Communication*, Ablex Publishing, USA.

Turnbull, D. 2007. "Maps Narratives and Trails: Performativity, Hodology and distributed Knowledges in Complex Adaptive Systems – an Approach to Emergent Mapping." *Geographical Research* 45:2, 140-149.

Warren, P.S. Katti, M. Ermann, M. and 2006. "Urban bioacoustics: it's not just noise." *Animal Behaviour* 71, pp. 491–502.

Chapter 15

Multi-Modal Mapping Methods And Methodologies

Laurene Vaughan[1]

William Cartwright[2]

Sébastien Caquard[3]

[1] School of Media and Communication and Research Leader within the Design Research Institute at RMIT University, Melbourne
[2] School of Mathematical and Geospatial Sciences, RMIT University, Melbourne
[3] Department of Geography, Planning and Environment, Concordia University, Montréal

The Interdisciplinary Essence of Mapping

Cartography has always embraced different ways of representing geography. Whilst the general perception is that cartography is just about 'maps', other representations of geography have been explored as adjuncts to, or substitutes for, 'the map'. This concept is not new. If we look at the second edition of the textbook on cartography by Erwin Raisz – General Cartography - he states in the Preface:

> *"The public, confronted with geographical problems of all parts of the world, demanded maps, and maps were wanting. Our old types of maps and atlases failed to present the true nature, life, economics, and problems of various parts of the world. In response to increased demands, a new school of cartography is developing with great promise for a renaissance of this art."*
>
> (Raisz, 1948, p. vi)

It is interesting that Raisz saw great promise for 'this art' when he referred to the advancement of cartography to address the ways in which we should represent the true nature of geography. He understood the scientific 'underpinnings' needed to properly map the earth, and the technology essential for producing, replicating and communicating this information. But, he also saw that 'art' was one of the key elements needed for cartographers to develop appropriate artefacts and for users to exploit them to better understand the Earth.

Science and technology has been embraced by cartography as a means to ensure that what is presented is scientifically 'correct' – products are considered to 'work' if they are scientifically 'elegant', technologically 'buildable' and 'deliverable' using contemporary communications systems. But science or technology, need not always take on the primary roles of mapping. There exists the need to address the role that art can take to facilitate further development of contemporary cartography. This is especially the case in areas where non-traditional methods of representation can be applied to promote the building of representations, visualizations and tools that support seeing geography differently, thus complementing more 'mainstream' cartographic products.

Contemporary methods for depicting the earth and its cultural and natural attributes use graphic and non-graphic formats, maps and map-related artefacts, to visualize geography and to build virtual landscapes and environments. The discipline area of cartography, traditionally, has applied art, science and technology to map-making in order to design and realise these products. Prior to the mid 1950s cartographic artefacts were built under the theoretical and practical 'umbrella' of this partnership of art, science and technology. However, since then, it can be said that the theory and methodology associated with visualizing geography has focused more on science and technology, and away from art. This 'move' away from art was accelerated by: 1) computing, computers and complete automated systems; and 2) a 'quest' to gain 'scientific legitimacy' by using Scientific Visualization, applied through Geographical Visualization theories and products, as a lodestone for gauging the 'quality' of cartographic theories and subsequent developed cartographic applications.

Much of the research conducted by the International Cartographic Association's Commission on Art and Cartography has explored how alternative cartographic information design and contemporary communication systems can facilitate the provision of a tool for geographical interpretation, complementing the mathematically and scientifically precise methods used to develop mainstream mapping products, thus producing a more balanced 'suite' of artefacts that facilitate the provision of many different, and complementary, interpretations and representations of geography. The provision of 'different' tools allow for improved visualization of geography and therefore a better understanding of the essential components of the mapped world.

The impetus for the workshop *'Mapping Environmental Issues in the City'* was the desire by the convenors (two cartographers and one designer) to explore the similarities and differences that exist across different disciplinary approaches to mapping. Previous explorations and discussions by members of the Arts and Cartography Working Group in the International Cartographic Association had focussed on different aesthetic and cultural approaches to mapping. The key points of connection and comparison in these events had been the shared interest and the desire to map and represent space and the cultural production of place. Although many interesting presentations and publications have emerged, and lively conversations have been had, as a result there continued to be a sense that something core was missing. As members of this Working Group, we sought to know more about the synergies and contradictions that may exist across the arts, science and humanities in relation to mapping: but how could we explore such a thing? How could we facilitate an opportunity and context where we could look closely at epistemologies and methods, and reflect on the various map forms within some framework of consistency? The answer to our questioning led us to identify that one set of data was needed, housed in a shared database this would be the mechanism for such a challenge. Consequently a database of one place (Montréal) that focused on some particular features or concerns (environmental conditions) was developed as the locus for this exploration by the participants various mapping practices. The aim was to use this one common element as a means to more clearly consider diversity.

As is recorded in the chapters of this text, the outcomes of this workshop are as diverse as the people that participated. Even when there are overlaps in disciplinary expertise the approach to the mapping place and to engaging with the database has emerged from individual areas of concern. The outcomes of this shared investigation evidence the rich history of cartography and its alignment to broader methods of communication and representation of place. They are multi-dimensional and multi-sensorial and extend well beyond the limitations of the two-dimensional orthographic map. Such a rich approach to the representation of place is of course not new, nor does it lie outside of the traditions of cartography and geography. There is a long history of cartographic representation using a diversity of representational forms; maps have been spoken, performed, drawn; they have mixed truth and fantasy, been represented as complete and unfinished. And so, within such a rich trajectory of knowing, how did the outcomes of this interdisciplinary investigation of one place add to our understanding of the multiple ways that we map space?

Common Data and Multiple Perspectives

When discussing the practises of mapping from multiple disciplinary perspectives, there is a point in the discussion when it becomes apparent that even though the same words are being used, the meaning of those words is significantly different across the domains. A simple term such as 'map', which can be both a verb and a noun is understood quite differently by an artist, an architect, a geographer or digital visualiser. In the course of the workshop it became apparent that although the data and database had been devised as a common or neutral element, there were in fact multiple understandings and perceptions of what these were and how they relate to the real or lived city of Montréal. To explore the multiple perspectives in greater detail, the workshop participants participated in a series of round table discussions where they explored the four questions:

- What is data?
- How is data represented?
- How do conceptual models inform database design?
- What is the relationship between data the database and the real world?

These questions were explored for two hours in round table conversations (Figure 15.1). For some the answers seemed logical, simple and a conventional aspect of their professional activities. For others the concepts and issues were quite foreign, and their comprehension of things such as the relationship between data and databases were complex and contrary to others in their discussion groups. A full transcript of the outcomes of the workshop was posted on the workshop blog page (http://mappingworkshop.wordpress.com/), and the following is a summation of what emerged.

Figure 15.1 Debating the data during the workshop.

Data?

It was generally agreed that we are in a time of data overload and data gluttony. New technologies have made it possible to collect and collate a multitude of information about people, places and things and that these are compiled as data that can be used for analysis and representation. Consequently the quality of the data that is available is variable, spanning from being high and reliable, and that which is of low quality, variable, unreliable and often unscientific.

Data Representation

The fraught nature of data representation was a common understanding across the workshop participants. There was consensus amongst the group regarding individual concerns about the ability to accurately represent data and all that is associated with it. Thereby accepting that data has lived meaning for those that it is drawn from, and then finding means for representing it with integrity is a challenge in the mapping process, this resulted in a level of scepticism regarding people's ability to really authentically represent any data and any place in all its richness. The over simplification of complex data through stylised graphic or textual means was another area of associated concern. Raising questions regarding the influence that a data interpreter and map maker can have on the real meaning of data and how it may be manipulated in the mapping process for reasons such as aesthetic preferences, or size and scale limitations of the mapping outcome. There was some expectation that digital technologies were having a positive influence on this, but at the same time the digital divide exists and the issues of access and equity continue to impact on both the making of maps and their use.

Conceptual Models

It was generally understood that data is 'dumb' in that it is what it is, and that each bit is a singular yet interconnected entity, while it is the database that brings the data to life and makes it accessible. The conceptual model that underpins the database design is a significant factor that is the first level of interpretation of the dumb data, and it is what frames our ability to learn from it. The value of database becomes evident through the use of it, and it is through this use that the values of the database become evident. What we can and can't do, and what is included and excluded, are the essential elements in once again being able to use data and ultimately to engage with the data that is housed within the database. Yet at the same time, the database will only ever be as good as the data and metadata that it contains. A database is a relative entity, and it is the relationships, which the database enables which are amongst the measures of its success.

Data – Database – Real World

There is an intricate relationship between the data, the database and the real world. Data is drawn from the real world, it is the world broken down into small or specific bits. The database is the means and the conduit for the data to be brought back into the real world, for it to be transformed and reconfigured into a new form that can be viewed, critiqued and seen in relation to other bits, and new narratives of interrelational knowing to be formed.

These narrative alternatives are a dynamic element of the database, the ability of the database to reconfigure and present back to the user new or unexpected relationships between the bits can enable us to re-conceive that which is known. The database can serve as disruptive device within the norms of the everyday, which may also result in deeper knowing even when the knowing is limited to the quantifiable bits of data. The relationship between the quantifiable and qualitative data was raised as an issue in the design and limitations of a database. Raising questions regarding how a database that is designed to tell us information about the environmental issues of a city can actually tell us of what it is like to live there? It is in this way that links between interpretation and representation become essential to considerations of the authority of a database and what is produced from it.

Figure 15.2 Multiple perspectives and shared understandings.

As the workshop evolved and the various participants presented their responses to the database it was evident that the issue of inhabiting the data of the database was a core area of concerns and interest (Fig 2). Whether this was through community participation, the adaptation of new technologies or linking to cultural production in the local community, discovering or interpreting the strategies of people who live, work and commute within the landscape of the data in place, was the common element.

The Challenge of Representing Place

Just as it is understood that there are multiple forms of mapping practices and maps exist in multiple forms, there is also a shared understanding of the limitations of maps to adequately represent the richness of the data that they draw on. This has resulted in a plethora of publications, projects and exhibitions that investigate the complexity, diversity and limitations. From Elsewhere Mapping (Abrams & Hall 2006), Maps that Lie (Monmonier 1996) to Atlas's of Emotion (2002), the European Novel (Moretti 1998), and of Spatial Practices (Basar 2006), or this recent series of publications by Springer exploring the relationship between Art and Cartography (Cartwright, Gartner & Lehn 2009) or the Mapping of Different Geographies (Kriz, Cartwright & Hurni 2010) is just to name a few of those that have been delving in this area. What becomes evident as one begins to delve is that the concerns of those who participated in the workshop exist across the broader domain. Thus raising the question of why? Why is there such an interest in exploring multiple forms of maps, what is the seduction of the map and the act of representing place?

The answer to these questions may lay in the multifarious nature of place itself. Place is a subjective entity, realised through the interconnections between the individual, their social relations, the environment and the meaning that these have for them (Sack 1997). If place is such an individual and subjective entity, then any representation of place will be unique to both the map maker and the map that is the outcome. Although there is an ideal that map making is an objective practice, where the impartial data that is the basis of the practice ensures that the outcome will be as well, this may not be the case. To suggest that maps are subjective is to suggest that they are inaccurate or lack the authority that they are entrusted with, but this is not necessarily the case. For what is the authority that we hope for from a map? A map like any designed artefact or system is designed with an end use in mind. A map of a road system is designed to show us where the roads are so that we can find our way from one place to another. A map of environmental water conditions in a city, tell us about the environment and can be used to help make links to other information regarding health, water management etc. Maps are multifarious entities. They are designed with an intended use in mind and the databases that underpin the map design are the basis of this. Like all design artefacts maps have limitations; they may not tell us all we hope to know, and they may be used in unintentional ways.

New technologies provide opportunities for us to rethink the workings of a map; what they can represent and what they can communicate. Innovations in database design and technology are an important aspect of these new representations, resulting in what Nadia Amoroso names as 'datascapes' (2010 p. 68). These datascapes provide an opportunity to include both the visual and non-visual factors that are essential to the mapping and representation of place. The projects and essays that have been discussed in this book are perhaps best articulated as datascapes; but unlike the digital basis of Amoroso's discussion, this body of work integrates analogue and digital technologies, and many adopt a performative methodology.

This investigation of the 'Database of the Environmental Issues of the City' as realised in the city of Montréal, has exposed a database that not only records the factors of the life of a city, but is also a lived and living database. Where the actions of those who engage with the database realise it meaning, through a rich mix of diverse representations of place.

References:

Abrams, J., & Hall, P. 2006. *Else/Where: Mapping New Cartographies of Networks and Territories*, University of Minnesota Design Institute, Minnesota

Amoroso, N. 2010. *The exposed city: mapping the urban invisibles*, Routledge, New York.

Basar, S. 2006. *Did Someone Say Participate?: An Atlas of Spatial Practice*, The MIT Press, Cambridge, Mass.

Bruno, G. 2002. *Atlas of Emotion: Journeys in Art, Architecture, and Film,* Verso, London.

Cartwright, W., Gartner, G. & Lehn, A. 2009. *Cartography and Art*, Lecture Notes in Geoinformation and Cartography, Spinger-Verlag, Berlin, Heidelberg.

Kriz, K., Cartwright, W. & Hurni, L. 2010. *Mapping Different Geographies*, Lecture Notes in Geoinformation and Cartography, Spinger-Verlag, Berlin, Heidelberg.

Monmonier, M. 1996. *How to Lie with Maps*, University of Chicago Press, Chicago

Moretti, F. 1998. *Atlas of the European Novel*, Verso, London.

Raisz, E. 1948. *General Cartography*. Second edition, McGraw-Hill Book Company Inc, New York.

Sack, R. D. 1997. *Homo Geographicus*, John Hopkins University Press, Baltimore.